CAMBRIDGE LIBRARY COLLECTION

Books of enduring scholarly value

Botany and Horticulture

Until the nineteenth century, the investigation of natural phenomena, plants and animals was considered either the preserve of elite scholars or a pastime for the leisured upper classes. As increasing academic rigour and systematisation was brought to the study of 'natural history', its subdisciplines were adopted into university curricula, and learned societies (such as the Royal Horticultural Society, founded in 1804) were established to support research in these areas. A related development was strong enthusiasm for exotic garden plants, which resulted in plant collecting expeditions to every corner of the globe, sometimes with tragic consequences. This series includes accounts of some of those expeditions, detailed reference works on the flora of different regions, and practical advice for amateur and professional gardeners.

Old Time Gardens

The American social historian and antiquarian Alice Morse Earle (1851–1911) published this work in 1901. She was a prolific writer of books and pamphlets on pre-revolutionary New England, and her writings were very popular with readers who took great interest in the social history and material culture of their country. In this work, which contains more than 200 illustrations, Earle describes the historic and modern gardens of the north-eastern seaboard, the gardening activities – for pleasure as well as for food – of early settlers, and the progress of plant-hunters and nursery-men such as John Bartram in discovering and categorising new specimens, as well as the introduction into the United States of cottage garden favourites from Europe and exotica from the Far East. Earle's *Sundials and Roses of Yesterday* (1902) is also reissued in this series.

Cambridge University Press has long been a pioneer in the reissuing of out-of-print titles from its own backlist, producing digital reprints of books that are still sought after by scholars and students but could not be reprinted economically using traditional technology. The Cambridge Library Collection extends this activity to a wider range of books which are still of importance to researchers and professionals, either for the source material they contain, or as landmarks in the history of their academic discipline.

Drawing from the world-renowned collections in the Cambridge University Library and other partner libraries, and guided by the advice of experts in each subject area, Cambridge University Press is using state-of-the-art scanning machines in its own Printing House to capture the content of each book selected for inclusion. The files are processed to give a consistently clear, crisp image, and the books finished to the high quality standard for which the Press is recognised around the world. The latest print-on-demand technology ensures that the books will remain available indefinitely, and that orders for single or multiple copies can quickly be supplied.

The Cambridge Library Collection brings back to life books of enduring scholarly value (including out-of-copyright works originally issued by other publishers) across a wide range of disciplines in the humanities and social sciences and in science and technology.

Old Time Gardens
Newly Set Forth

A Book of the Sweet o' the Year

Alice Morse Earle

CAMBRIDGE
UNIVERSITY PRESS

University Printing House, Cambridge, CB2 8BS, United Kingdom

Cambridge University Press is part of the University of Cambridge.
It furthers the University's mission by disseminating knowledge in the pursuit of education, learning and research at the highest international levels of excellence.

www.cambridge.org
Information on this title: www.cambridge.org/9781108076616

This edition first published 1902
This digitally printed version 2015

ISBN 978-1-108-07661-6 Paperback

Selected books of related interest, also reissued in the CAMBRIDGE LIBRARY COLLECTION

Amherst, Alicia: *A History of Gardening in England* (1895) [ISBN 9781108062084]

Anonymous: *The Book of Garden Management* (1871) [ISBN 9781108049399]

Blaikie, Thomas: *Diary of a Scotch Gardener at the French Court at the End of the Eighteenth Century* (1931) [ISBN 9781108055611]

Candolle, Alphonse de: *The Origin of Cultivated Plants* (1886) [ISBN 9781108038904]

Drewitt, Frederic Dawtrey: *The Romance of the Apothecaries' Garden at Chelsea* (1928) [ISBN 9781108015875]

Evelyn, John: *Sylva, Or, a Discourse of Forest Trees* (2 vols., fourth edition, 1908) [ISBN 9781108055284]

Farrer, Reginald John: *In a Yorkshire Garden* (1909) [ISBN 9781108037228]

Field, Henry: *Memoirs of the Botanic Garden at Chelsea* (1878) [ISBN 9781108037488]

Forsyth, William: *A Treatise on the Culture and Management of Fruit-Trees* (1802) [ISBN 9781108037471]

Haggard, H. Rider: *A Gardener's Year* (1905) [ISBN 9781108044455]

Hibberd, Shirley: *Rustic Adornments for Homes of Taste* (1856) [ISBN 9781108037174]

Hibberd, Shirley: *The Amateur's Flower Garden* (1871) [ISBN 9781108055345]

Hibberd, Shirley: *The Fern Garden* (1869) [ISBN 9781108037181]

Hibberd, Shirley: *The Rose Book* (1864) [ISBN 9781108045384]

Hogg, Robert: *The British Pomology* (1851) [ISBN 9781108039444]

Hogg, Robert: *The Fruit Manual* (1860) [ISBN 9781108039451]

Hooker, Joseph Dalton: *Kew Gardens* (1858) [ISBN 9781108065450]

Jackson, Benjamin Daydon: *Catalogue of Plants Cultivated in the Garden of John Gerard, in the Years 1596–1599* (1876) [ISBN 9781108037150]

Jekyll, Gertrude: *Home and Garden* (1900) [ISBN 9781108037204]

Jekyll, Gertrude: *Wood and Garden* (1899) [ISBN 9781108037198]

Johnson, George William: *A History of English Gardening, Chronological, Biographical, Literary, and Critical* (1829) [ISBN 9781108037136]

Knight, Thomas Andrew: *A Selection from the Physiological and Horticultural Papers Published in the Transactions of the Royal and Horticultural Societies* (1841) [ISBN 9781108037297]

Lindley, John: *The Theory of Horticulture* (1840) [ISBN 9781108037242]

Loudon, Jane: *Instructions in Gardening for Ladies* (1840) [ISBN 9781108055659]

Mollison, John: *The New Practical Window Gardener* (1877) [ISBN 9781108061704]

Paris, John Ayrton: *A Biographical Sketch of the Late William George Maton M.D.* (1838) [ISBN 9781108038157]

Paxton, Joseph, and Lindley, John: *Paxton's Flower Garden* (3 vols., 1850–3) [ISBN 9781108037280]

Repton, Humphry and Loudon, John Claudius: *The Landscape Gardening and Landscape Architecture of the Late Humphry Repton, Esq.* (1840) [ISBN 9781108066174]

Robinson, William: *The English Flower Garden* (1883) [ISBN 9781108037129]

Robinson, William: *The Subtropical Garden* (1871) [ISBN 9781108037112]

Robinson, William: *The Wild Garden* (1870) [ISBN 9781108037105]

Sedding, John D.: *Garden-Craft Old and New* (1891) [ISBN 9781108037143]

Veitch, James Herbert: *Hortus Veitchii* (1906) [ISBN 9781108037365]

Ward, Nathaniel: *On the Growth of Plants in Closely Glazed Cases* (1842) [ISBN 9781108061131]

For a complete list of titles in the Cambridge Library Collection please visit:
www.cambridge.org/features/CambridgeLibraryCollection/books.htm

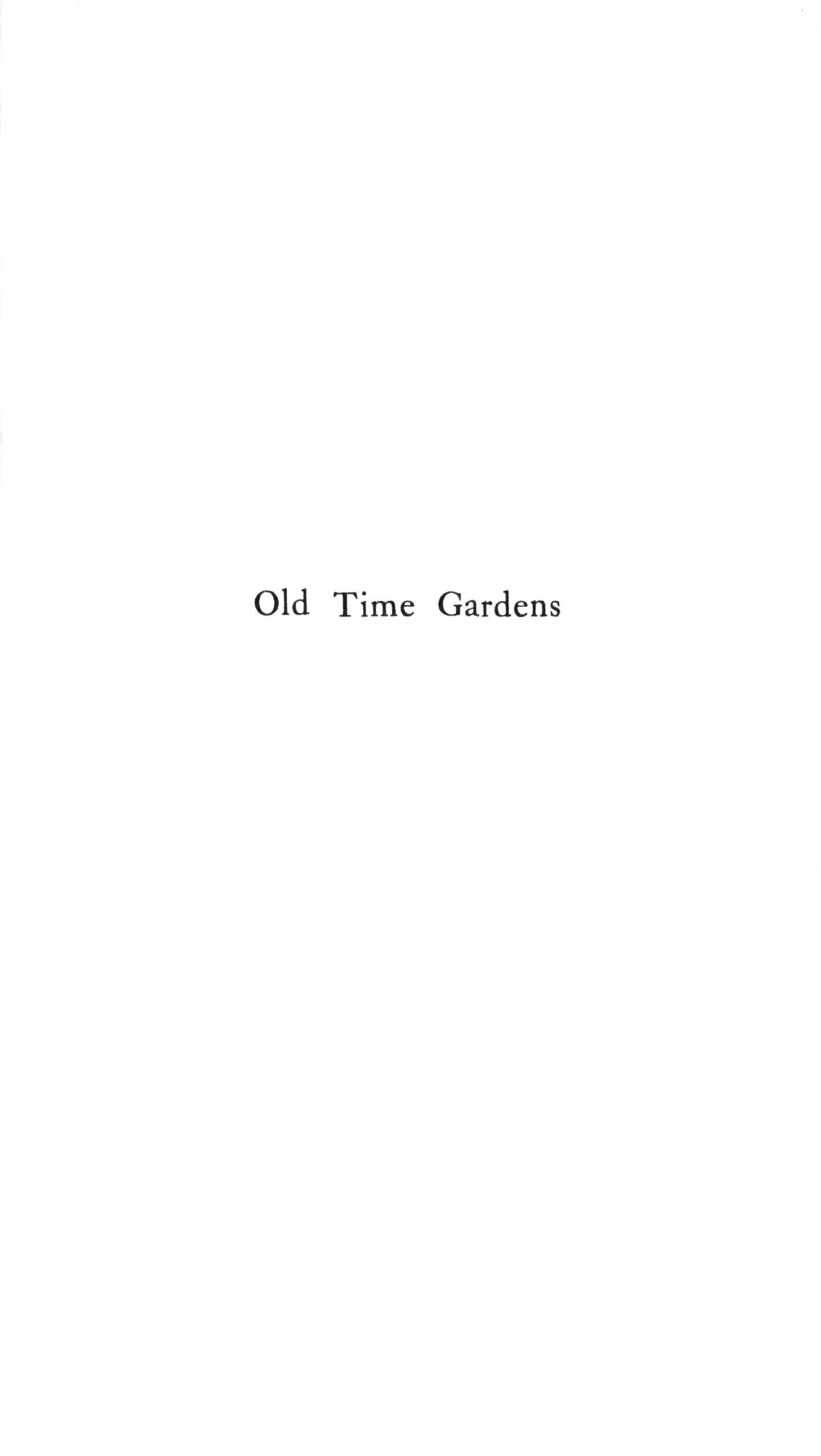

Old Time Gardens

·The MM Co·

OLD·TIME GARDENS

Newly set forth

by

ALICE MORSE EARLE

A BOOK OF

THE SWEET O THE YEAR

"Life is sweet, brother! There's day and night, brother! both sweet things: sun, moon and stars, brother! all sweet things: There is likewise a wind on the heath"

NEW YORK
THE MACMILLAN COMPANY
LONDON MACMILLAN & CO LTD
MCMII

Set up and electrotyped November, 1901. Reprinted December, 1901.

Norwood Press
J. S. Cushing & Co.—Berwick & Smith
Norwood, Mass., U.S.A.

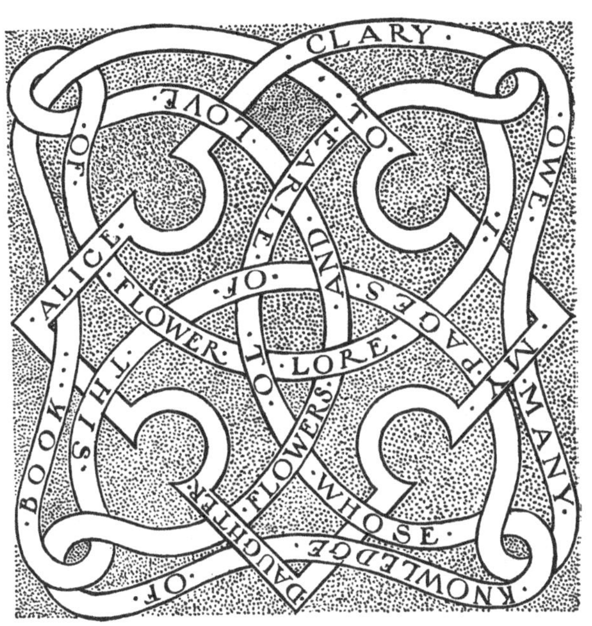

CLARY
LOVE
OF
TO
EARLE
OWE
I
AND
ALICE
OF
FLOWER
PAGES
TO
LORE
MY
THIS
BOOK
FLOWERS
WHOSE
MANY
DAUGHTER
KNOWLEDGE
OF

Contents

List of Illustrations

The end papers of this book bear a design of the flower Ambrosia.

The vignette on the title-page is re-drawn from one in *The Compleat Body of Husbandry*, Thomas Hale, 1756. It represents "Love laying out the surface of the earth in a garden."

The device of the dedication is an ancient garden-knot for flowers, from *A New Orchard and Garden*, William Lawson, 1608.

The chapter initials are from old wood-cut initials in the English Herbals of Gerarde, Parkinson, and Cole.

Old Time Gardens

Old Time Gardens

CHAPTER I

COLONIAL GARDEN-MAKING

"There is not a softer trait to be found in the character of those stern men than that they should have been sensible of these flower-roots clinging among the fibres of their rugged hearts, and felt the necessity of bringing them over sea, and making them hereditary in the new land."

— *American Note-book*, NATHANIEL HAWTHORNE.

FTER ten wearisome weeks of travel across an unknown sea, to an equally unknown world, the group of Puritan men and women who were the founders of Boston neared their Land of Promise; and their noble leader, John Winthrop, wrote in his Journal that "we had now fair Sunshine Weather and so pleasant a sweet Aire as did much refresh us, and there came a smell off the Shore like the Smell of a Garden."

A *Smell of a Garden* was the first welcome to our ancestors from their new home; and a pleasant and perfect emblem it was of the life that awaited them.

They were not to become hunters and rovers, not to be eager to explore quickly the vast wilds beyond; they were to settle down in the most domestic of lives, as tillers of the soil, as makers of gardens.

What must that sweet air from the land have been to the sea-weary Puritan women on shipboard, laden to them with its promise of a garden! for I doubt not every woman bore with her across seas some little package of seeds and bulbs from her English home garden, and perhaps a tiny slip or plant of some endeared flower; watered each day, I fear, with many tears, as well as from the surprisingly scant water supply which we know was on board that ship.

And there also came flying to the *Arbella* as to the Ark, a Dove — a bird of promise — and soon the ship came to anchor.

> "With hearts revived in conceit new Lands and Trees they spy,
> Scenting the Cædars and Sweet Fern from heat's reflection dry,"

wrote one colonist of that arrival, in his *Good Newes from New England.* I like to think that Sweet Fern, the characteristic wild perfume of New England, was wafted out to greet them. And then all went on shore in the sunshine of that ineffable time and season, — a New England day in June, — and they "gathered store of fine strawberries," just as their Salem friends had on a June day on the preceding year gathered strawberries and "sweet Single Roses" so resembling the English Eglantine that the hearts of the women must have ached within them with fresh homesickness. And ere long all had dwelling-places, were they but humble log cabins; and pasture lands and commons were portioned out; and in a short time all had garden-plots, and thus, with sheltering roof-trees, and warm firesides, and with gardens, even in this lonely new world, they had *homes*. The first entry in the Plymouth Records is a significant one; it is the assignment of "Meresteads and Garden-Plotes," not meresteads alone, which were farm lands, but home gardens: the outlines of these can still be seen in Plymouth town. And soon all sojourners who bore news back to England of the New-Englishmen and New-Englishwomen, told of ample store of gardens. Ere a year had passed hopeful John Winthrop wrote, "My Deare Wife, wee are here in a Paradise." In four years the chronicler Wood said in his *New England's Prospect*, "There is growing here all manner of herbs for meat and medicine, and that not only in planted gardens, but in the woods, without the act and help of man." Governor Endicott had by that time a very creditable garden.

And by every humble dwelling the homesick goodwife or dame, trying to create a semblance of her fair English home so far away, planted in her "garden plot" seeds and roots of homely English flowers and herbs, that quickly grew and blossomed and smiled on bleak New England's rocky shores as sturdily and happily as they had bloomed in the old gardens and by the ancient door sides in England. What good cheer they must have brought! how they must have been beloved! for these old English garden flowers are such gracious things;

marvels of scent, lavish of bloom, bearing such genial faces, growing so readily and hardily, spreading so quickly, responding so gratefully to such little care: what pure refreshment they bore in their blossoms, what comfort in their seeds; they must have seemed an emblem of hope, a promise of a new and happy home. I rejoice over every one that I know was in those little colonial gardens, for each one added just so much measure of solace to what seems to me, as I think upon it, one of the loneliest, most fearsome things that gentlewomen ever had to do, all the harder because neither by poverty nor by unavoidable stress were they forced to it; they came across-seas willingly, for conscience' sake. These women were not accustomed to the thought of emigration, as are European folk to-day; they had no friends to greet them in the new land; they were to encounter wild animals and wild men; sea and country were unknown — they could scarce expect ever to return: they left everything, and took nothing of comfort but their Bibles and their flower seeds. So when I see one of the old English flowers, grown of those days, blooming now in my garden, from the unbroken chain of blossom to seed of nearly three centuries, I thank the flower for all that its forbears did to comfort my forbears, and I cherish it with added tenderness.

We should have scant notion of the gardens of these New England colonists in the seventeenth century were it not for a cheerful traveller named John Josselyn, a man of everyday tastes and much inquisitiveness, and the pleasing literary style which

Garden of the Johnson Mansion, Germantown, Pennsylvania.

comes from directness, and an absence of self-consciousness. He published in 1672 a book entitled *New England's Rarities discovered*, etc., and in 1674 another volume giving an account of his two voyages hither in 1638 and 1663. He made a very careful list of vegetables which he found thriving in the new land; and since his flower list is the earliest known, I will transcribe it in full; it isn't long, but there is enough in it to make it a suggestive outline which we can fill in from what we know of the plants to-day, and form a very fair picture of those gardens.

"Spearmint,
Rew, will hardly grow
Fetherfew prospereth exceedingly;
Southernwood, is no Plant for this Country, Nor
Rosemary. Nor
Bayes.
White-Satten groweth pretty well, so doth
Lavender-Cotton. But
Lavender is not for the Climate.
Penny Royal
Smalledge.
Ground Ivey, or Ale Hoof.
Gilly Flowers will continue two Years.
Fennel must be taken up, and kept in a Warm Cellar all Winter
Horseleek prospereth notably
Holly hocks
Enula Canpana, in two years time the Roots rot.
Comferie, with White Flowers.
Coriander, and
Dill, and

Annis thrive exceedingly, but Annis Seed, as also the seed of Fennel seldom come to maturity; the Seed of Annis is commonly eaten with a Fly.
Clary never lasts but one Summer, the Roots rot with the Frost.
Sparagus thrives exceedingly, so does
Garden Sorrel, and
Sweet Bryer or Eglantine
Bloodwort but sorrily, but
Patience and
English Roses very pleasantly.
Celandine, by the West Country now called Kenning Wort grows but slowly.
Muschater, as well as in England
Dittander or Pepperwort flourisheth notably and so doth Tansie."

These lists were published fifty years after the landing of the Pilgrims at Plymouth; from them we find that the country was just as well stocked with vegetables as it was a hundred years later when other travellers made lists, but the flowers seem few; still, such as they were, they formed a goodly sight. With rows of Hollyhocks glowing against the rude stone walls and rail fences of their little yards; with clumps of Lavender Cotton and Honesty and Gillyflowers blossoming freely; with Feverfew "prospering" to sow and slip and pot and give to neighbors just as New England women have done with Feverfew every year of the centuries that have followed; with "a Rose looking in at the window" — a Sweetbrier, Eglantine, or English Rose — these colonial dames might well find "Patience growing very pleasantly" in their hearts as in their gardens.

They had plenty of pot herbs for their accustomed savoring; and plenty of medicinal herbs for their

Garden at Grumblethorp, Germantown, Pennsylvania.

wonted dosing. Shakespeare's "nose-herbs" were not lacking. Doubtless they soon added to these garden flowers many of our beautiful native blooms, rejoicing if they resembled any beloved English flowers, and quickly giving them, as we know, familiar old English plant-names.

And there were other garden inhabitants, as truly English as were the cherished flowers, the old garden weeds, which quickly found a home and thrived in triumph in the new soil. Perhaps the weed seeds came over in the flower-pot that held a sheltered plant or cutting; perhaps a few were mixed with garden seeds; perhaps they were in the straw or other packing of household goods: no one knew the manner of their coming, but there they were, Motherwort, Groundsel, Chickweed, and Wild Mustard, Mullein and Nettle, Henbane and Wormwood. Many a goodwife must have gazed in despair at the persistent Plantain, "the Englishman's foot," which seems to have landed in Plymouth from the Mayflower.

Josselyn made other lists of plants which he found in America, under these headings: —

"Such plants as are common with us in England.
Such plants as are proper to the Country.
Such plants as are proper to the Country and have no name.
Such plants as have sprung up since the English planted, and kept cattle in New England."

In these lists he gives a surprising number of English weeds which had thriven and rejoiced in their new home.

Mr. Tuckerman calls Josselyn's list of the fishes of the new world a poor makeshift; his various lists of plants are better, but they are the lists of an herbalist, not of a botanist. He had some acquaintance with the practice of physic, of which he narrates some examples; and an interest in kitchen recipes, and included a few in his books. He said that Parkinson or another botanist might have "found in

Garden of the Bartram House, Philadelphia, Pennsylvania.

New England a thousand, at least, of plants never heard of nor seen by any Englishman before," and adds that he was himself an indifferent observer. He certainly lost an extraordinary opportunity of distinguishing himself, indeed of immortalizing himself; and it is surprising that he was so heedless, for Englishmen of that day were in general eager botanists. The study of plants was new, and was deemed of such absorbing interest and fascination that some rigid Puritans feared they might lose their immortal souls through making their new plants their idols.

When Josselyn wrote, but few of our American flowers were known to European botanists; Indian

Garden of Abigail Adams.

Corn, Pitcher Plant, Columbine, Milkweed, Everlasting, and Arbor-vitæ had been described in printed books, and the Evening Primrose. A history of Canadian and other new plants, by Dr. Cornuti, had been printed in Europe, giving thirty-seven of our plants; and all English naturalists were longing to add to the list; the ships which brought over homely seeds and plants for the gardens of the colonists carried back rare American seeds and plants for English physic gardens.

In Pennsylvania, from the first years of the settlement, William Penn encouraged his Quaker followers to plant English flowers and fruit in abundance, and to try the fruits of the new world. Father Pastorius, in his Germantown settlement, assigned to each family a garden-plot of three acres, as befitted a man who left behind him at his death a manuscript poem of many thousand words on the pleasures of gardening, the description of flowers, and keeping of bees. George Fox, the founder of the Friends, or Quakers, died in 1690. He had travelled in the colonies; and in his will he left sixteen acres of land to the Quaker meeting in the city of Philadelphia. Of these sixteen acres, ten were for "a close to put Friends' horses in when they came afar to the Meeting, that they may not be Lost in the Woods," while the other six were for a site for a meeting-house and school-house, and "for a Playground for the Children of the town to Play on, and for a Garden to plant with Physical Plants, for Lads and Lasses to know Simples, and to learn to make Oils and Ointments." Few as are these words, they convey a positive picture of Fox's intent, and a pleasing picture it is. He had seen what interest had been awakened and what instruction conveyed through the "Physick-Garden" at Chelsea, England; and he promised to himself similar interest and information from the study of plants and flowers by the

Quaker "lads and lasses" of the new world. Though nothing came from this bequest, there was a later fulfilment of Fox's hopes in the establishment of a successful botanic garden in Philadelphia, and, in the planting, growth, and flourishing in the province of Pennsylvania of the loveliest gardens in the new world; there floriculture reached by the time of the Revolution a very high point; and many exquisite gardens bore ample testimony to the "pride of life," as well as to the good taste and love of flowers of Philadelphia Friends. The garden at Grumblethorp, the home of Charles J. Wister, Esq., of Germantown, Pennsylvania, shown on page 7, dates to colonial days and is still flourishing and beautiful.

In 1728 was established, by John Bartram, in Philadelphia, the first botanic garden in America. The ground on which it was planted, and the stone dwelling-house he built thereon in 1731, are now part of the park system of Philadelphia. A view of the garden as now in cultivation is given on page 9. Bartram travelled much in America, and through his constant correspondence and flower exchanges with distinguished botanists and plant growers in Europe, many native American plants became well known in foreign gardens, among them the Lady's Slipper and Rhododendron. He was a Quaker, — a quaint and picturesque figure, — and his example helped to establish the many fine gardens in the vicinity of Philadelphia. The example and precept of Washington also had important influence; for he was constant in his desire and his effort to secure every good and new plant, grain,

Garden at Mount Vernon-on-the-Potomac. Home of George Washington.

shrub, and tree for his home at Mount Vernon. A beautiful tribute to his good taste and that of his wife still exists in the Mount Vernon flower garden, which in shape, Box edgings, and many details is precisely as it was in their day. A view of its well-ordered charms is shown opposite page 12. Whenever I walk in this garden I am deeply grateful to the devoted women who keep it in such perfection, as an object-lesson to us of the dignity, comeliness, and beauty of a garden of the olden times.

There is little evidence that a general love and cultivation of flowers was as common in humble homes in the Southern colonies as in New England and the Middle provinces. The teeming abundance near the tropics rendered any special gardening unnecessary for poor folk; flowers grew and blossomed lavishly everywhere without any coaxing or care. On splendid estates there were splendid gardens, which have nearly all suffered by the devastations of war — in some towns they were thrice thus scourged. So great was the beauty of these Southern gardens and so vast the love they provoked in their owners, that in more than one case the life of the garden's master was merged in that of the garden. The British soldiers during the War of the Revolution wantonly destroyed the exquisite flowers at "The Grove," just outside the city of Charleston, and their owner, Mr. Gibbes, dropped dead in grief at the sight of the waste.

The great wealth of the Southern planters, their constant and extravagant following of English customs and fashions, their fertile soil and favorable climate, and their many slaves, all contributed to the successful making of elaborate gardens. Even as early as 1682 South Carolina gardens were declared to be "adorned with such Flowers as to the Smell or Eye are pleasing or agreeable, as the Rose, Tulip, Lily, Carnation, &c." William Byrd wrote of the terraced gardens of Virginia homes. Charleston dames vied with each other in the beauty of their gardens, and Mrs. Logan, when seventy years old, in 1779, wrote a treatise called *The Gardener's Kalendar*. Eliza Lucas Pinckney of Charleston was devoted to practical floriculture and horticulture. Her introduction of indigo raising into South Carolina revolutionized the trade products of the state and brought to it vast wealth. Like many other women and many men of wealth and culture at that time, she kept up a constant exchange of letters, seeds, plants, and bulbs with English people of like tastes. She received from them valuable English seeds and shrubs; and in turn she sent to England what were so eagerly sought by English flower raisers, our native plants. The good will and national pride of ship captains were enlisted; even young trees of considerable size were set in hogsheads, and transported, and cared for during the long voyage.

The garden at Mount Vernon is probably the oldest in Virginia still in original shape. In Maryland are several fine, formal gardens which do not date, however, to colonial days; the beautiful one at Hampton, the home of the Ridgelys, in Baltimore County, is shown on pages 57, 60 and 95. In both North and South Carolina the gardens were exquisite. Many were laid out by competent landscape gardeners, and were kept in order by skilled workmen, negro slaves, who were carefully trained from childhood to special labor, such

Gate and Hedge of Preston Garden.

as topiary work. In Camden and Charleston the gardens vied with the finest English manor-house gardens. Remains of their beauty exist, despite devastating wars and earthquakes. Views of the Preston Garden, Columbia, South Carolina, are shown on pages 15 and 18 and facing page 54. They are now the grounds of the Presbyterian College for Women. The hedges have been much reduced within a few years; but the garden still bears a surprising resemblance to the Garden of the Generalife, Granada. The Spanish garden has fewer flowers and more fountains, yet I think it must have been the model for the Preston Garden. The climax of magnificence in Southern gardens has been for years, at Magnolia-on-the-Ashley, the ancestral home of the Draytons since 1671. It is impossible to describe the affluence of color in this garden in springtime; masses of unbroken bloom on giant Magnolias; vast Camellia Japonicas, looking, leaf and flower, thoroughly artificial, as if made of solid wax; splendid Crape Myrtles, those strange flower-trees; mammoth Rhododendrons; Azaleas of every Azalea color,—all surrounded by walls of the golden Banksia Roses, and hedges covered with Jasmine and Honeysuckle. The Azaleas are the special glory of the garden; the bushes are fifteen to twenty feet in height, and fifty or sixty feet in circumference, with rich blossoms running over and crowding down on the ground as if color had been poured over the bushes; they extend in vistas of vivid hues as far as the eye can reach. All this gay and brilliant color is overhung by a startling contrast, the most sombre and gloomy thing in nature, great Live-oaks heavily draped with gray Moss; the avenue of largest Oaks was planted two centuries ago.

I give no picture of this Drayton Garden, for a photograph of these many acres of solid bloom is a meaningless thing. Even an oil painting of it is confused and disappointing. In the garden itself the excess of color is as cloying as its surfeit of scent pouring from the thousands of open flower cups; we long for green hedges, even for scanter bloom and for fainter fragrance. It is not a garden to live in, as are our box-bordered gardens of the North, our cheerful cottage borders, and our well-balanced Italian gardens, so restful to the eye; it is a garden to look at and wonder at.

The Dutch settlers brought their love of flowering bulbs, and the bulbs also, to the new world. Adrian Van der Donck, a gossiping visitor to New Netherland when the little town of New Amsterdam had about a thousand inhabitants, described the fine kitchen gardens, the vegetables and fruits, and gave an interesting list of garden flowers which he found under cultivation by the Dutch vrouws. He says:

"Of the Flowers. The flowers in general which the Netherlanders have introduced there are the white and red roses of different kinds, the cornelian roses, and stock roses; and those of which there were none before in the country, such as eglantine, several kinds of gillyflowers, jenoffelins, different varieties of fine tulips, crown imperials, white lilies, the lily frutularia, anemones, baredames, violets, marigolds, summer sots, etc. The clove tree has also been introduced, and there are various indigenous trees that bear handsome flowers, which are unknown in the Netherlands. We also find there some flowers of native growth, as, for instance, sunflowers, red and yellow lilies, mountain lilies, morning stars, red, white, and yellow maritoffles (a very sweet flower), several species of bell flowers, etc., to which I have not given particular attention, but *amateurs* would hold them in high estimation and make them widely known."

I wish I knew what a Cornelian Rose was, and Jenoffelins, Baredames, and Summer Sots; and what the Lilies were and the Maritoffles and Bell Flowers. They all sound so cheerful and homelike

Fountain Path in Preston Garden, Columbia, South Carolina.

—just as if they bloomed well. Perhaps the Cornelian Rose may have been striped red and white like cornelian stone, and like our York and Lancaster Rose.

Tulips are on all seed and plant lists of colonial days, and they were doubtless in every home dooryard in New Netherland. Governor Peter Stuyvesant had a fine farm on the Bouwerie, and is said to have had a flower garden there and at his home, White Hall, at the Battery, for he had forty or fifty negro slaves who were kept at work on his estate. In the city of New York many fine formal gardens lingered, on what are now our most crowded streets, till within the memory of persons now living. One is described as full of "Paus bloemen of all hues, Laylocks, and tall May Roses and Snowballs intermixed with choice vegetables and herbs all bounded and hemmed in by huge rows of neatly-clipped Box-edgings."

An evidence of increase in garden luxury in New York is found in the advertisement of one Theophilus Hardenbrook, in 1750, a practical surveyor and architect, who had an evening school for teaching architecture. He designed pavilions, summer-houses, and garden seats, and "Green-houses for the preservation of Herbs with winding Funnels through the walls so as to keep them warm." A picture of the green-house of James Beekman, of New York, 1764, still exists, a primitive little affair. The first glass-house in North America is believed to be one built in Boston for Andrew Faneuil, who died in 1737.

Mrs. Anne Grant, writing of her life near Albany in the middle of the eighteenth century, gives a very good description of the Schuyler garden. Skulls of domestic animals on fence posts, would seem astounding had I not read of similar decorations in old Continental gardens. Vines grew over these grisly fence-capitals and birds built their nests in them, so in time the Dutch housewife's peaceful kitchen garden ceased to resemble the kraal of an African chieftain; to this day, in South Africa, natives and Dutch Boers thus set up on gate posts the skulls of cattle.

Mrs. Grant writes of the Dutch in Albany: —

"The care of plants, such as needed peculiar care or skill to rear them, was the female province. Every one in town or country had a garden. Into this garden no foot of man intruded after it was dug in the Spring. I think I see yet what I have so often beheld — a respectable mistress of a family going out to her garden, on an April morning, with her great calash, her little painted basket of seeds, and her rake over her shoulder to her garden of labours. A woman in very easy circumstances and abundantly gentle in form and manners would sow and plant and rake incessantly."

We have happily a beautiful example of the old Dutch manor garden, at Van Cortlandt Manor, at Croton-on-Hudson, New York, still in the possession of the Van Cortlandt family. It is one of the few gardens in America that date really to colonial days. The manor house was built in 1681; it is one of those fine old Dutch homesteads of which we still have many existing throughout New York, in which dignity, comfort, and fitness are so happily combined. These homes are, in the words of a traveller of colonial days, "so pleasant in their building, and contrived so delightful." Above all, they are so suited to their surroundings that they seem an intrinsic part of the landscape, as they do of the old life of this Hudson River Valley.

Door in Wall of Kitchen Garden at Van Cortlandt Manor.

I do not doubt that this Van Cortlandt garden was laid out when the house was built; much of it must be two centuries old. It has been extended, not altered; and the grass-covered bank supporting the upper garden was replaced by a brick terrace wall about sixty years ago. Its present form dates to the days when New York was a province. The upper garden is laid out in formal flower beds; the lower border is rich in old vines and shrubs, and all the beloved old-time hardy plants. There is in the manor-house an ancient portrait of the child Pierre Van Cortlandt, painted about the year 1732. He stands by a table bearing a vase filled with old garden flowers — Tulip, Convolvulus, Harebell, Rose, Peony, Narcissus, and Flowering Almond; and it is the pleasure of the present mistress of the manor, to see that the garden still holds all the great-grandfather's flowers.

There is a vine-embowered old door in the wall under the piazza (see opposite page 20) which opens into the kitchen and fruit garden; a wall-door so quaint and old-timey that I always remind me of Shakespeare's lines in *Measure for Measure:* —

> "He hath a garden circummured with brick,
> Whose western side is with a Vineyard back'd;
> And to that Vineyard is a planchéd gate
> That makes his opening with this bigger key:
> The other doth command a little door
> Which from the Vineyard to the garden leads."

The long path is a beautiful feature of this garden (it is shown in the picture of the garden opposite page 24); it dates certainly to the middle of the eighteenth century. Pierre Van Cortlandt, the son of the child with the vase of flowers, and grandfather of the present generation bearing his surname, was born in 1762. He well recalled playing along this garden path when he was a child; and that one day he and his little sister Ann (Mrs. Philip Van Rensselaer) ran a race along this path and through the garden to see who could first "see the baby" and greet their sister, Mrs. Beekman, who came riding to the manor-house up the hill from Tarrytown, and through the avenue, which shows on the right-hand side of the garden-picture. This beautiful young woman was famed everywhere for her grace and loveliness, and later equally so for her intelligence and goodness, and the prominent part she bore in the War of the Revolution. She was seated on a pillion behind her husband, and she carried proudly in her arms her first baby (afterward Dr. Beekman) wrapped in a scarlet cloak. This is one of the home-pictures that the old garden holds. Would we could paint it!

In this garden, near the house, is a never failing spring and well. The house was purposely built near it, in those days of sudden attacks by Indians; it has proved a fountain of perpetual youth for the old Locust tree, which shades it; a tree more ancient than house or garden, serene and beautiful in its hearty old age. Glimpses of this manor-house garden and its flowers are shown on many pages of this book, but they cannot reveal its beauty as a whole — its fine proportions, its noble background, its splendid trees, its turf, its beds of bloom. Oh! how beautiful a garden can be, when for two hundred years it has been loved and cherished, ever nurtured, ever guarded; how plainly it shows such care!

Another Dutch garden is pictured opposite page 32, the garden of the Bergen Homestead, at Bay Ridge, Long Island. Let me quote part of its description, written by Mrs. Tunis Bergen:—

"Over the half-open Dutch door you look through the vines that climb about the stoop, as into a vista of the past. Beyond the garden is the great Quince orchard of hundreds of trees in pink and white glory. This orchard has a story which you must pause in the garden to hear. In the Library at Washington is preserved, in quaint manuscript, 'The Battle of Brooklyn,' a farce written and said to have been performed during the British occupation. The scene is partly laid in 'the orchard of one Bergen,' where the British hid their horses after the battle of Long Island—this is the orchard; but the blossoming Quince trees tell no tale of past carnage. At one side of the garden is a quaint little building with moss-grown roof and climbing hop-vine—the last slave kitchen left standing in New York—on the other side are rows of homely beehives. The old Locust tree overshadowing is an ancient landmark—it was standing in 1690. For some years it has worn a chain to bind its aged limbs together. All this beauty of tree and flower lived till 1890, when it was swept away by the growing city. Though now but a memory, it has the perfume of its past flowers about it."

The Locust was so often a "home tree" and so fitting a one, that I have grown to associate ever with these Dutch homesteads a light-leaved Locust tree, shedding its beautiful flickering shadows on the long roof. I wonder whether there was any association or tradition that made the Locust the house-friend in old New York!

The first nurseryman in the new world was stern old Governor Endicott of Salem. In 1644 he wrote to Governor Winthrop, "My children burnt mee at least 500 trees by setting the ground on fire neere them" — which was a very pretty piece of mischief for sober Puritan children. We find all thoughtful men of influence and prominence in all the colonies raising various fruits, and selling trees and plants, but they had no independent business nurseries.

If tradition be true, it is to Governor Endicott we owe an indelible dye on the landscape of eastern Massachusetts in midsummer. The Dyer's-weed or Woad-waxen (*Genista tinctoria*), which, in July, covers hundreds of acres in Lynn, Salem, Swampscott, and Beverly with its solid growth and brilliant yellow bloom, is said to have been brought to this country as the packing of some of the governor's household belongings. It is far more probable that he brought it here to raise it in his garden for dyeing purposes, with intent to benefit the colony, as he did other useful seeds and plants. Woad-waxen, or Broom, is a persistent thing; it needs scythe, plough, hoe, and bitter labor to eradicate it. I cannot call it a weed, for it has seized only poor rock-filled land, good for naught else; and the radiant beauty of the Salem landscape for many

Garden at Van Cortlandt Manor.

weeks makes us forgive its persistence, and thank Endicott for bringing it here.

> "The Broom,
> Full-flowered and visible on every steep,
> Along the copses runs in veins of gold."

The Broom flower is the emblem of mid-summer, the hottest yellow flower I know—it seems to throw out heat. I recall the first time I saw it growing; I was told that it was "Salem Wood-wax." I had heard of "Roxbury Waxwork," the Bitter-sweet, but this was a new name, as it was a new tint of yellow, and soon I had its history, for I find Salem people rather proud both of the flower and its story.

Oxeye Daisies (Whiteweed) are also by vague tradition the children of Governor Endicott's planting. I think it far more probable that they were planted and cherished by the wives of the colonists, when their beloved English Daisies were found unsuited to New England's climate and soil. We note the Woad-waxen and Whiteweed as crowding usurpers, not only because they are persistent, but because their great expanses of striking bloom will not let us forget them. Many other English plants are just as determined intruders, but their modest dress permits them to slip in comparatively unobserved.

It has ever been characteristic of the British colonist to carry with him to any new home the flowers of old England and Scotland, and characteristic of these British flowers to monopolize the earth. Sweetbrier is called "the missionary-plant," by the Maoris in New Zealand, and is there regarded as a tiresome weed, spreading and holding the ground. Some homesick missionary or his more homesick wife bore it there; and her love of the home plant impressed even the savage native. We all know the story of the Scotch settlers who carried their beloved Thistles to Tasmania "to make it seem like home," and how they lived to regret it. Vancouver's Island is completely overrun with Broom and wild Roses from England.

The first commercial nursery in America, in the sense of the term as we now employ it, was established about 1730 by Robert Prince, in Flushing, Long Island, a community chiefly of French Huguenot settlers, who brought to the new world many French fruits by seed and cuttings, and also a love of horticulture. For over a century and a quarter these Prince Nurseries were the leading ones in America. The sale of fruit trees was increased in 1774 (as we learn from advertisements in the *New York Mercury* of that year), by the sale of "Carolina Magnolia flower trees, the most beautiful trees that grow in America, and 50 large Catalpa flower trees; they are nine feet high to the under part of the top and thick as one's leg," also other flowering trees and shrubs.

The fine house built on the nursery grounds by William Prince suffered little during the Revolution. It was occupied by Washington and afterwards house and nursery were preserved from depredations by a guard placed by General Howe when the British took possession of Flushing. Of course, domestic nursery business waned in time of war; but an excellent demand for American shrubs and trees sprung up among the officers of the British army, to send home to gardens in England and Germany. Many an English garden still has ancient plants and trees from the Prince Nurseries.

The "Linnæan Botanic Garden and Nurseries" and the "Old American Nursery" thrived once more at the close of the war, and William Prince the second entered in charge; one of his earliest ventures of importance was the introduction of Lombardy Poplars. In 1798 he advertises ten thousand trees, ten to seventeen feet in height. These became the most popular tree in America, the emblem of democracy—and a warmly hated tree as well. The eighty acres of nursery grounds were a centre of botanic and horticultural interest for the entire country; every tree, shrub, vine, and plant known to England and America was eagerly sought for; here the important botanical treasures of Lewis and Clark found a home. William Prince wrote several notable horticultural treatises; and even his trade catalogues were prized. He established the first steamboats between Flushing and New York, built roads and bridges on Long Island, and was a public-spirited, generous citizen as well as a man of science. His son, William Robert Prince, who died in 1869, was the last to keep up the nurseries, which he did as a scientific rather than a commercial establishment. He botanized the entire length of the Atlantic States with Dr. Torrey, and sought for collections of trees and wild flowers in California with the same eagerness that others there sought gold. He was a devoted promoter of the native silk industry, having vast plantations of Mulberries in many cities; for one at Norfolk, Virginia, he was offered $100,000. It is a curious fact that the interest in Mulberry culture and the practice of its cultivation was so uni-

Garden at Prince Homestead, Flushing, Long Island.

versal in his neighborhood (about the year 1830), that cuttings of the Chinese Mulberry (*Morus multicaulis*) were used as currency in all the stores in the vicinity of Flushing, at the rate of 12½ cents each.

The Prince homestead, a fine old mansion, is here shown; it is still standing, surrounded by that forlorn sight, a forgotten garden. This is of considerable extent, and evidences of its past dignity appear in the hedges and edgings of Box; one symmetrical great Box tree is fifty feet in circumference. Flowering shrubs, unkempt of shape, bloom and beautify the waste borders each spring, as do the oldest Chinese Magnolias in the United States. Gingkos, Paulownias, and weeping trees, which need no gardener's care, also flourish and are of unusual size. There are some splendid evergreens, such as Mt. Atlas Cedars; and the oldest and finest Cedar of Lebanon in the United States. It seemed sad, as I looked at the evidences of so much past beauty and present decay, that this historic house and garden should not be preserved for New York, as the house and garden of John Bartram, the Philadelphia botanist, have been for his native city.

While there are few direct records of American gardens in the eighteenth century, we have many instructing side glimpses through old business letter-books. We find Sir Harry Frankland ordering Daffodils and Tulips for the garden he made for Agnes Surriage; and it is said that the first Lilacs ever seen in Hopkinton were planted by him for her. The gay young nobleman and the lovely woman are in the dust, and of all the beautiful things belonging to them there remain a splendid Portuguese fan, which stands as a memorial of that tragic crisis in their life — the great Lisbon earthquake; and the Lilacs, which still mark the site of her house and blossom each spring as a memorial of the shadowed romance of her life in New England.

Let me give two pages from old letters to illustrate what I mean by side glimpses at the contents of colonial gardens. The fine Hancock mansion in Boston had a carefully-filled garden long previous to the Revolution. Such letters as the following were sent by Mr. Hancock to England to secure flowers for it: —

"My Trees and Seeds for Capt. Bennett Came Safe to Hand and I like them very well. I Return you my hearty Thanks for the Plumb Tree and Tulip Roots you were pleased to make me a Present off, which are very Acceptable to me. I have Sent my friend Mr. Wilks a mmo. to procure for me 2 or 3 Doz. Yew Trees, Some Hollys and Jessamine Vines, and if you have Any Particular Curious Things not of a high Price, will Beautifye a flower Garden Send a Sample with the Price or a Catalogue of 'em, I do not intend to spare Any Cost or Pains in making my Gardens Beautifull or Profitable.

"P.S. The Tulip Roots you were Pleased to make a present off to me are all Dead as well."

We find Richard Stockton writing in 1766 from England to his wife at their beautiful home "Morven," in Princeton, New Jersey: —

"I am making you a charming collection of bulbous roots, which shall be sent over as soon as the prospect of freezing on your coast is over. The first of April, I believe, will be time enough for you to put them in your sweet little flower garden, which you so fondly cultivate. Suppose I inform you that I design a ride to Twickenham the latter end of next month principally to view Mr. Pope's gardens and grotto, which I am told remain nearly as he left them; and that I shall take with me a gentleman who draws well, to lay down an exact plan of the whole."

The fine line of Catalpa trees set out by Richard Stockton, along the front of his lawn, were in full flower when he rode up to his house on a memorable July day to tell his wife that he had signed the Declaration of American Independence. Since then Catalpa trees bear everywhere in that vicinity

Old Box at Prince Homestead.

the name of Independence trees, and are believed to be ever in bloom on July 4th.

In the delightful diary and letters of Eliza Southgate Bowne (*A Girl's Life Eighty Years Ago*), are other side glimpses of the beautiful gardens of old Salem, among them those of the wealthy merchants of the Derby family. Terraces and arches show a formality of arrangement, for they were laid out by a Dutch gardener whose descendants still live in Salem. All had summer-houses, which were larger and more important buildings than what are to-day termed summer-houses; these latter were known in Salem and throughout Virginia as bowers. One summer-house had an arch through it with three doors on each side which opened into little apartments; one of them had a staircase by which you could ascend into a large upper room, which was the whole size of the building. This was constructed to command a fine view, and was ornamented with Chinese articles of varied interest and value; it was used for tea-drinkings. At the end of the garden, concealed by a dense Weeping Willow, was a thatched hermitage, containing the life-size figure of a man reading a prayer-book; a bed of straw and some broken furniture completed the picture. This was an English fashion, seen at one time in many old English gardens, and held to be most romantic. Apparently summer evenings were spent by the Derby household and their visitors wholly in the garden and summer-house. The diary keeper writes naïvely, "The moon shines brighter in this garden than anywhere else."

The shrewd and capable women of the colonies who entered so freely and successfully into business ventures found the selling of flower seeds a congenial occupation, and often added it to the pursuit of other callings. I think it must have been very pleasant to buy packages of flower seed at the same time and place where you bought your best bonnet,

Old Dutch Garden of Bergen Homestead.

and have all sent home in a bandbox together; each would prove a memorial of the other; and long after the glory of the bonnet had departed, and the bonnet itself was ashes, the thriving Sweet Peas and Larkspur would recall its becoming charms. I have often seen the advertisements of these seedswomen in old newspapers; unfortunately they seldom gave printed lists of their store of seeds. Here is one list printed in a Boston newspaper on March 30, 1760:—

Lavender.
Palma Christi.
Cerinthe or Honeywort, loved of bees.
Tricolor.
Indian Pink.
Scarlet Cacalia.
Yellow Sultans.
Lemon African Marigold.
Sensitive Plants.
White Lupine.
Love Lies Bleeding.
Patagonian Cucumber.
Lobelia.
Catchfly.
Wing-peas.
Convolvulus.
Strawberry Spinage.
Branching Larkspur.
White Chrysanthemum.
Nigaella Romano.
Rose Campion.
Snap Dragon.
Nolana prostrata.
Summer Savory.
Hyssop.
Red Hawkweed.
Red and White Lavater.
Scarlet Lupine.
Large blue Lupine.
Snuff flower.
Caterpillars.
Cape Marigold.
Rose Lupine.
Sweet Peas.
Venus' Navelwort.
Yellow Chrysanthemum.
Cyanus minor.
Tall Holyhock.
French Marigold.
Carnation Poppy.
Globe Amaranthus.
Yellow Lupine.
Indian Branching Coxcombs.
Iceplants.

Thyme.
Sweet Marjoram.
Tree Mallows.
Everlasting.
Greek Valerian.
Tree Primrose.
Canterbury Bells.
Purple Stock.
Sweet Scabiouse.
Columbine.
Pleasant-eyed Pink.
Dwarf Mountain Pink.
Sweet Rocket.
Horn Poppy.
French Honeysuckle.
Bloody Wallflower.
Sweet William.
Honesty (to be sold in small parcels that every one may have a little).
Persicaria.
Polyanthos.
50 Different Sorts of mixed Tulip Roots.
Ranunculus
Gladiolus.
Starry Scabiouse.
Curled Mallows.
Painted Lady topknot peas.
Colchicum.
Persian Iris.
Star Bethlehem.

This list is certainly a pleasing one. It gives opportunity for flower borders of varied growth and rich color. There is a quality of some minds which may be termed historical imagination. It is the power of shaping from a few simple words or details of the faraway past, an ample picture, full of light and life, of which these meagre details are but a framework. Having this list of the names of these sturdy old annuals and perennials, what do you perceive besides the printed words? I see that the old mid-century garden where these seeds found a home was a cheerful place from earliest spring to autumn; that it had many bulbs, and thereafter a constant succession of warm blooms till the Coxcombs, Marigolds, Colchicums and Chrysanthemums yielded to New England's frosts. I know that the garden had beehives and that the bees were loved; for when they sallied out of their straw bee-skepes, these happy bees found their favorite blossoms planted to welcome them: Cerinthe, dropping with honey; Cacalia, a sister flower; Lupine, Larkspur, Sweet Marjoram, and Thyme — I can

Old Garden at Duck Cove Farm in Narragansett.

taste the Thyme-scented classic honey from that garden! There was variety of foliage as well as bloom, the dovelike Lavender, the glaucous Horned Poppy, the glistening Iceplants, the dusty Rose Campion.

Stately plants grew from the little seed-packets; Hollyhocks, Valerian, Canterbury Bells, Tree Primroses looked down on the low-growing herbs of the border; and there were vines of Convolvulus and Honeysuckle. It was a garden overhung by clouds of perfume from Thyme, Lavender, Sweet Peas, Pleasant-eyed Pink, and Stock. The garden's mistress looked well after her household; ample store of savory pot herbs grow among the finer blossoms.

It was a garden for children to play in. I can see them; little boys with their hair tied in queues, in knee breeches and flapped coats like their stately fathers, running races down the garden path, as did the Van Cortlandt children; and demure little girls in caps and sacques and aprons, sitting in cubby houses under the Lilac bushes. I know what flowers they played with and how they played, for they were my great-grandmothers and grandfathers, and they played exactly what I did, and sang what I did when I was a child in a garden. And suddenly my picture expands, as a glow of patriotic interest thrills me in the thought that in this garden were sheltered and amused the boys of one hundred and forty years ago, who became the heroes of our American Revolution; and the girls who were Daughters of Liberty, who spun and wove and knit for their soldiers, and drank heroically their miserable Liberty tea. I fear the garden faded when bitter war scourged the land, when the women turned from their flower beds to the plough and the field, since their brothers and husbands were on the frontier.

But when that winter of gloom to our country and darkness to the garden was ended, the flowers bloomed still more brightly, and to the cheerful seedlings of the old garden is now given perpetual youth and beauty; they are fated never to grow faded or neglected or sad, but to live and blossom and smile forever in the sunshine of our hearts through the magic power of a few printed words in a time-worn old news-sheet.

CHAPTER II

FRONT DOORYARDS

"There are few of us who cannot remember a front yard garden which seemed to us a very paradise in childhood. Whether the house was a fine one and the enclosure spacious, or whether it was a small house with only a narrow bit of ground in front, the yard was kept with care, and was different from the rest of the land altogether. . . . People do not know what they lose when they make way with the reserve, the separateness, the sanctity, of the front yard of their grandmothers. It is like writing down family secrets for any one to read; it is like having everybody call you by your first name, or sitting in any pew in church."

— *Country Byways*, SARAH ORNE JEWETT, 1881.

LD New England villages and small towns and well-kept New England farms had universally a simple and pleasing form of garden called the front yard or front dooryard. A few still may be seen in conservative communities in the New England states and in New York or Pennsylvania. I saw flourishing ones this summer in Gloucester, Marblehead, and Ipswich. Even where the front yard was but a narrow strip of land before a tiny cottage, it was carefully fenced in, with a gate that was kept rigidly closed and latched. There seemed to be a law which shaped and bounded the front yard; the side fences extended from the corners of the house to the front fence on the edge of the road, and thus formed naturally the guarded parallelogram. Often the fence around the front yard was the only one on the farm; everywhere else were boundaries of great stone walls; or if there were rail

The Flowering Almond under the Window.

fences, the front yard fence was the only painted one. I cannot doubt that the first gardens that our foremothers had, which were wholly of flowering plants, were front yards, little enclosures hard won from the forest.

The word yard, not generally applied now to any enclosure of elegant cultivation, comes from the same root as the word garden. Garth is another derivative, and the word exists much disguised in orchard. In the sixteenth century yard was used in formal literature instead of garden; and later Burns writes of " Eden's bonnie yard, Where yeuthful lovers first were pair'd."

This front yard was an English fashion derived from the forecourt so strongly advised by Gervayse Markham (an interesting old English writer on floriculture and husbandry), and found in front of many a yeoman's house, and many a more pretentious house as well in Markham's day. Forecourts were common in England until the middle of the eighteenth century, and may still be seen. The forecourt gave privacy to the house even when in the centre of a town. Its readoption is advised with handsome dwellings in England, where ground-space is limited, — and why not in America, too?

The front yard was sacred to the best beloved, or at any rate the most honored, garden flowers of the house mistress, and was preserved by its fences from inroads of cattle, which then wandered at their will and were not housed, or even enclosed at night. The flowers were often of scant variety, but were those deemed the gentlefolk of the flower world. There was a clump of Daffodils and of the Poet's Narcissus in early spring, and stately Crown Imperial; usually, too, a few scarlet and yellow single Tulips, and Grape Hyacinths. Later came Phlox in abundance — the only native American plant,— Canterbury Bells, and ample and glowing London Pride. Of course there were great plants of white and blue Day Lilies, with their beautiful and decorative leaves, and purple and yellow Flower de Luce. A few old-fashioned shrubs always were seen. By inflexible law there must be a Lilac, which might be the aristocratic Persian Lilac. A Syringa, a flowering Currant, or Strawberry bush made sweet the front yard in spring, and sent wafts of fragrance into

Peter's Wreath.

the house-windows. Spindling, rusty Snowberry bushes were by the gate, and Snowballs also, or our native Viburnums. Old as they seem, the Spiræas and Deutzias came to us in the nineteenth century from Japan; as did the flowering Quinces and Cherries. The pink Flowering Almond dates back to the oldest front yards (see page 39), and Peter's Wreath certainly seems an old settler and is found now in many front yards that remain. The lovely full-flowered shrub of Peter's Wreath, on page 41, which was photographed for this book, was all that remained of a once-loved front yard.

The glory of the front yard was the old-fashioned early red "Piny," cultivated since the days of Pliny. I hear people speaking of it with contempt as a vulgar flower, — flaunting is the conventional derogatory adjective, — but I glory in its flaunting. The modern varieties, of every tint from white through flesh color, coral, pink, ruby color, salmon, and even yellow, to deep red, are as beautiful as Roses. Some are sweet-scented; and they have no thorns, and their foliage is ever perfect, so I am sure the Rose is jealous.

I am as fond of the Peony as are the Chinese, among whom it is flower queen. It is by them regarded as an aristocratic flower; and in old New England towns fine Peony plants in an old garden are a pretty good indication of the residence of what Dr. Holmes called New England Brahmins. In Salem and Portsmouth are old "Pinys" that have a hundred blossoms at a time — a glorious sight. A Japanese name is "Flower-of-prosperity"; another name, "Plant-of-twenty-days," because its glories last during that period of time.

Rhododendrons are to the modern garden what the Peony was in the old-fashioned flower border; and I am glad the modern flower cannot drive the old one out. They are equally varied in coloring, but the Peony is a much hardier plant, and I like it far better. It has no blights, no bugs, no dis-

Peonies in a Salem Garden.

eases, no running out, no funguses; it doesn't have to be covered in winter, and it will bloom in the shade. No old-time or modern garden is to me fully furnished without Peonies; see how fair they are in this Salem garden. I would grow them in some corner of the garden for their splendid healthy foliage if they hadn't a blossom. The *Pæonia tenuifolia* in particular has exquisite feathery foliage. The great Tree Peony, which came from China, grows eight feet or more in height, and is a triumph of the flower world; but it was not known to the oldest front yards. Some of the Tree Peonies have finely displayed leafage of a curious and very gratifying tint of green. Miss Jekyll, with her usual felicity, compares its blue cast with pinkish shading to the vari-colored metal alloys of the Japanese bronze workers — a striking comparison. The single Peonies of recent years are of great beauty, and will soon be esteemed here as in China.

Not the least of the Peony's charms is its exceeding trimness and cleanliness. The plants always look like a well-dressed, well-shod, well-gloved girl of birth, breeding, and of equal good taste and good health; a girl who can swim, and skate, and ride, and play golf. Every inch has a well-set, neat, cared-for look which the shape and growth of the plant keeps from seeming artificial or finicky. See the white Peony on page 44; is it not a seemly, comely thing, as well as a beautiful one?

No flower can be set in our garden of more distinct antiquity than the Peony; the Greeks believed it to be of divine origin. A green arbor of the fourteenth century in England is described as set around with Gillyflower, Tansy, Gromwell, and "Pyonys powdered ay betwene" — just as I like to see Peonies set to this day, "powdered"

White Peonies.

everywhere between all the other flowers of the border.

I am pleased to note of the common flowers of the New England front yard, that they are no new things; they are nearly all Elizabethan of date — many are older still. Lord Bacon in his essay on gardens names many of them, Crocus, Tulip, Hyacinth, Daffodil, Flower de Luce, double Peony, Lilac, Lily of the Valley.

A favorite flower was the yellow garden Lily, the Lemon Lily, *Hemerocallis*, when it could be kept from spreading. Often its unbounded luxuriance exiled it from the front yard to the kitchen dooryard, as befell the clump shown facing page 48. Its pretty old-fashioned name was Liricon-fancy, given, I am told, in England to the Lily of the Valley. I know no more satisfying sight than a good bank of these Lemon Lilies in full flower. Below Flatbush there used to be a driveway leading to an old Dutch house, set at regular intervals with great clumps of Lemon Lilies, and their full bloom made them glorious. Their power of satisfactory adaptation in our modern formal garden is happily shown facing page 76, in the lovely garden of Charles E. Mather, Esq., in Haverford, Pennsylvania.

The time of fullest inflorescence of the nineteenth century front yard was when Phlox and Tiger Lilies bloomed; but the pinkish-orange colors of the latter (the oddest reds of any flower tints) blended most vilely and rampantly with the crimson-purple of the Phlox; and when London Pride joined with its glowing scarlet, the front yard fairly ached. Nevertheless, an adaptation of that front-yard bloom can be most effective in a garden border, when white Phlox only is planted, and the Tiger Lily or cultivated stalks of our wild nodding Lily rise above the white trusses of bloom. These wild Lilies grow very luxuriantly in the garden, often towering above our heads and forming great candelabra bearing two score or more blooms. It is no easy task to secure their deep-rooted rhizomes in the meadow. I know a young man who won his sweetheart by the patience and assiduity with which he dug for her all one broiling morning to secure for her the coveted Lily roots, and collapsed with mild sunstroke at the finish. Her gratitude and remorse were equal factors in his favor.

The Tiger Lily is usually thought upon as a truly old-fashioned flower, a veritable antique; it is a favorite of artists to place as an accessory in their colonial gardens, and of authors for their flower-beds of Revolutionary days, but it was not known either in formal garden or front yard, until after "the days when we lived under the King." The bulbs were first brought to England from Eastern Asia in 1804 by Captain Kirkpatrick of the East India Company's Service, and shared with the Japan Lily the honor of being the first Eastern Lilies introduced into European gardens. A few years ago an old gentleman, Mr. Isaac Pitman, who was then about eighty-five years of age, told me that he recalled distinctly when Tiger Lilies first appeared in our gardens, and where he first saw them growing in Boston. So instead of being an old-time flower, or even an old-comer from the Orient, it is one of the novelties of this century. How readily has it made itself at home, and even wandered wild down our roadsides!

The two simple colors of Phlox of the old-time front yard, white and crimson-purple, are now augmented by tints of salmon, vermilion, and rose. I recall with special pleasure the profuse garden decoration at East Hampton, Long Island, of a pure cherry-colored Phlox, generally a doubtful color to me, but there so associated with the white blooms of various other plants, and backed by a high hedge covered solidly with blossoming Honeysuckle, that it was wonderfully successful.

To other members of the Phlox family, all natives of our own continent, the old front yard owed much; the Moss Pink sometimes crowded out both Grass and its companion the Periwinkle; it is still found in our gardens, and bountifully also in our fields; either in white or pink, it is one of the satisfactions of spring, and its cheerful little blossom is of wonderful use in many waste places. An old-fashioned bloom, the low-growing *Phlox amœna*, with its queerly fuzzy leaves and bright crimson blossoms, was among the most distinctly old-fashioned flowers of the front yard. It was tolerated rather than cultivated, as was its companion, the Arabis or Rock Cress — both crowding, monopolizing creatures. I remember well how they spread over the beds and up the grass banks in my mother's garden, how sternly they were uprooted, in spite of the pretty name of the Arabis — "Snow in Summer."

Sometimes the front yard path had edgings of sweet single or lightly double white or tinted Pinks, which were not deemed as choice as Box edgings. Frequently large Box plants clipped into simple and natural shapes stood at the side of the doorstep, usually in the home of the well-to-do. A great shell might be on either side of the door-sill, if there chanced to be seafaring men-folk who lived or visited under the roof-tree. Annuals were few in number; sturdy old perennial plants of many years' growth were the most honored dwellers in the front yard, true representatives of old families. The Roses were few and poor, for there was usually some great tree just without the gate, an Elm or Larch, whose shadow fell far too near and heavily for the health of Roses. Sometimes there was a prickly semidouble yellow Rose, called by us a Scotch Rose, a Sweet Brier, or a rusty-flowered white Rose, similar, though inferior, to the Madame Plantier. A new fashion of trellises appeared in the front yard about sixty years ago, and crimson Boursault Roses climbed up them as if by magic.

One marked characteristic of the front yard was its lack of weeds; few sprung up, none came to seed-time; the enclosure was small, and it was a mark of good breeding to care for it well. Sometimes, however, the earth was covered closely under shrubs and plants with the cheerful little Ladies' Delights, and they blossomed in the chinks of the bricked path and under the Box edges. Ambrosia, too, grew everywhere, but these were welcome — they were not weeds.

Our old New England houses were suited in color and outline to their front yards as to our landscape. Lowell has given in verse a good description of the kind of New England house that always had a front dooryard of flowers.

Yellow Day Lilies.

"On a grass-green swell
That towards the south with sweet concessions fell,
It dwelt retired, and half had grown to be
As aboriginal as rock or tree.
It nestled close to earth, and seemed to brood
O'er homely thoughts in a half-conscious mood.
If paint it e'er had known, it knew no more
Than yellow lichens spattered thickly o'er
That soft lead gray, less dark beneath the eaves,
Which the slow brush of wind and weather leaves.
The ample roof sloped backward to the ground
And vassal lean-tos gathered thickly round,
Patched on, as sire or son had felt the need.
But the great chimney was the central thought.

* * * * *

It rose broad-shouldered, kindly, debonair,
Its warm breath whitening in the autumn air."

Sarah Orne Jewett, in the plaint of *A Mournful Villager*, has drawn a beautiful and sympathetic picture of these front yards, and she deplores their passing. I mourn them as I do every fenced-in or hedged-in garden enclosure. The sanctity and reserve of these front yards of our grandmothers was somewhat emblematic of woman's life of that day: it was restricted, and narrowed to a small outlook and monotonous likeness to her neighbor's; but it was a life easily satisfied with small pleasures, and it was comely and sheltered and carefully kept, and pleasant to the home household; and these were no mean things.

The front yard was never a garden of pleasure; children could not play in these precious little enclosed plots, and never could pick the flowers—

E

front yard and flowers were both too much respected. Only formal visitors entered therein, visitors who opened the gate and closed it carefully behind them, and knocked slowly with the brass knocker, and were ushered in through the ceremonious front door and the little ill-contrived entry, to the stiff fore-room or parlor. The parson and his wife entered that portal, and sometimes a solemn would-be sweetheart, or the guests at a tea party. It can be seen that every one who had enough social dignity to have a front door and a parlor, and visitors thereto, also desired a front yard with flowers as the external token of that honored standing. It was like owning a pew in church; you could be a Christian without having a pew, but not a respected one. Sometimes when there was a "vandue" in the house, reckless folk opened the front gate, and even tied it back. I attended one where the auctioneer boldly set the articles out through the windows under the Lilac bushes and even on the precious front yard plants. A vendue and a funeral were the only gatherings in country communities when the entire neighborhood came freely to an old homestead, when all were at liberty to enter the front dooryard. At the sad time when a funeral took place in the house, the front gate was fastened widely open, and solemn men-neighbors, in Sunday garments, stood rather uncomfortably and awkwardly around the front yard as the women passed into the house of mourning and were seated within. When the sad services began, the men too entered and stood stiffly by the door. Then through the front door, down the mossy path of the front yard, and through the open front gate was borne the master, the mistress, and then their children, and children's children. All are gone from our sight, many from our memory, and often too from our ken, while the Lilacs and Peonies and Flowers de Luce still blossom and flourish with perennial youth, and still claim us as friends.

At the side of the house or by the kitchen door would be seen many thrifty blooms: poles of Scarlet Runners, beds of Portulacas and Petunias, rows of Pinks, bunches of Marigolds, level expanses of Sweet Williams, banks of cheerful Nasturtiums, tangles of Morning-glories and long rows of stately Hollyhocks, which were much admired, but were seldom seen in the front yard, which was too shaded for them. Weeds grew here at the kitchen door in a rank profusion which was hard to conquer; but here the winter's Fuchsias or Geraniums stood in flower pots in the sunlight, and the tubs of Oleanders and Agapanthus Lilies.

The flowers of the front yard seemed to bear a more formal, a "company" aspect; conventionality rigidly bound them. Bachelor's Buttons might grow there by accident, but Marigolds never were tolerated, — they were pot herbs. Sunflowers were not even permitted in the flower beds at the side of the house unless these stretched down to the vegetable beds. Outside the front yard would be a rioting and cheerful growth of pink Bouncing Bet, or of purple Honesty, and tall straggling plants of a certain small flowered, ragged Campanula, and a white Mallow with flannelly leaves which, doubtless, aspired to inhabit the sacred bounds of the front yard (and probably dwelt there originally), and often were gladly permitted to grow in side gardens or kitchen dooryards, but which were regarded as interloping weeds by the guardians of the

Orange Day Lilies.

front yard, and sternly exiled. Sometimes a bed of these orange-tawny Day Lilies which had once been warmly welcomed from the Orient, and now were not wanted anywhere by any one, kept company with the Bouncing Bet, and stretched cheerfully down the roadside.

When the fences disappeared with the night rambles of the cows, the front yards gradually changed character; the tender blooms vanished, but the tall shrubs and the Peonies and Flower de Luce sturdily grew and blossomed, save where that dreary destroyer of a garden crept in — the desire for a lawn. The result was then a meagre expanse of poorly kept grass, with no variety, color, or change, — neither lawn nor front yard. It is ever a pleasure to me when driving in a village street or a country road to find one of these front yards still enclosed, or even to note in front of many houses the traces of a past front yard still plainly visible in the flourishing old-fashioned plants of many years' growth.

CHAPTER III

VARIED GARDENS FAIR

"And all without were walkes and alleys dight
With divers trees enrang'd in even rankes;
And here and there were pleasant arbors pight
And shadie seats, and sundry flowering bankes
To sit and rest the walkers wearie shankes."

— *Faerie Queene*, EDMUND SPENSER.

MANY simple forms of gardens were common besides the enclosed front yard; and as wealth poured in on the colonies, the beautiful gardens so much thought of in England were copied here, especially by wealthy merchants, as is noted in the first chapter of this book, and by the provincial governors and their little courts; the garden of Governor Hutchinson, in Milford, Massachusetts, is stately still and little changed.

English gardens, at the time of the settlement of America, had passed beyond the time when, as old Gervayse Markham said, "Of all the best Ornaments used in our English gardens, Knots and Mazes are the most ancient." A maze was a placing of low garden hedges of Privet, Box, or Hyssop, usually set in concentric circles which en-

Preston Garden.

closed paths, that opened into each other by such artful contrivance that it was difficult to find one's way in and out through these bewildering paths. "When well formed, of a man's height, your friend may perhaps wander in gathering berries as he cannot recover himself without your help."

The maze was not a thing of beauty, it was "nothing for sweetness and health," to use Lord Bacon's words; it was only a whimsical notion of gardening amusement, pleasing to a generation who liked to have hidden fountains in their gardens to sprinkle suddenly the unwary. I doubt if any mazes were ever laid out in America, though I have heard vague references to one in Virginia. Knots had been the choice adornment of the Tudor garden. They were not wholly a thing of the past when we had here our first gardens, and they have had a distinct influence on garden laying-out till our own day.

An Elizabethan poet wrote: —

> "My Garden sweet, enclosed with walles strong,
> Embanked with benches to sitt and take my rest;
> The knots so enknotted it cannot be expressed
> The arbores and alyes so pleasant and so dulce."

These garden knots were not flower beds edged with Box or Rosemary, with narrow walks between the edgings, as were the parterres of our later formal gardens. They were square, ornamental beds, each of which had a design set in some close-growing, trim plant, clipped flatly across the top, and the design filled in with colored earth or sand; and with no dividing paths. Elaborate models in complicated geometrical pattern were given in gardeners' books, for setting out these knots, which were first drawn on paper and subdivided into squares; then the square of earth was similarly divided, and set out by precise rules. William Lawson, the Izaak Walton of gardeners, gave, as a result of forty-eight years of experience, some very attractive directions for large "knottys" with different "thrids" of flowers, each of one color, which made the design appear as if "made of diverse colored ribands." One of his knots, from *A New Orchard and Garden* 1618, being a garden fashion in vogue when my forbears came to America, I have chosen as a device for the dedication of this book, thinking it, in Lawson's words, "so comely, and orderly placed, and so intermingled, that one looking thereon cannot but wonder." His knots had significant names, such as "Cinkfoyle; Flower de Luce; Trefoyle; Frette; Lozenge; Groseboowe; Diamond; Ovall; Maze."

Gervayse Markham gives various knot patterns to be bordered with Box cut eighteen inches broad at the bottom and kept flat at the top — with the ever present thought for the fine English linen. He has a varied list of circular, diamond-shaped, mixed, and "single impleated knots."

These garden knots were mildly sneered at by Lord Bacon; he said, "they be but toys, you see as good sights many times in tarts;" still I think they must have been quaint, and I should like to see a garden laid out to-day in these pretty Elizabethan knots, set in the old patterns, and with the old flowers. Nor did Parkinson and other practical gardeners look with favor on "curiously knotted

Box-edged Parterre at Hampton.

gardens," though all gave designs to "satisfy the desires" of their readers. "Open knots" were preferred; these were made with borders of lead, tiles, boards, or even the shankbones of sheep, "which will become white and prettily grace out the garden,"—a fashion I saw a few years ago around flower beds in Charlton, Massachusetts. "Round whitish pebble stones" for edgings were Parkinson's own invention, and proud he was of it, simple as it seems to us. These open knots were then filled in, but "thin and sparingly," with "English Flowers"; or with "Out-Landish Flowers," which were flowers fetched from foreign parts.

The parterre succeeded the knot, and has been used in gardens till the present day. Parterres were of different combinations, "well-contriv'd and ingenious." The "parterre of cut-work" was a Box-bordered formal flower garden, of which the garden at Hampton, Maryland (pages 57, 60, and 95), is a striking and perfect example; also the present garden at Mount Vernon (opposite page 12), wherein carefully designed flower beds, edged with Box, are planted with variety of flowers, and separated by paths. Sometimes, of old, fine white sand was carefully strewn on the earth under the flowers. The "parterre à l'Anglaise" had an elaborate design of vari-shaped beds edged with Box, but enclosing grass instead of flowers. In the "parterre de broderie" the Box-edged beds were filled with vari-colored earths and sands. Black earth could be made of iron filings; red earth of pounded tiles. This last-named parterre differed from a knot solely in having the paths among the beds. The *Retir'd Gard'ner* gives patterns for ten parterres.

The main walks which formed the basis of the garden design had in ancient days a singular name—forthrights; these were ever to be "spacious and fair," and neatly spread with colored sands or gravel. Parkinson says, "The fairer and larger your allies and walks be the more grace your garden should have, the lesse harm the herbes and flowers shall receive, and the better shall your weeders cleanse both the bed and the allies." "Covert-walks," or "shade-alleys," had trees meeting in an arch over them.

A curious term, found in references to old American flower beds and garden designs, as well as English ones, is the "goose-foot." A "goose-foot" consisted of three flower beds or three avenues radiating rather closely together from a small semicircle; and in some places and under some conditions it is still a charming and striking design, as you stand at the heel of the design and glance down the three avenues.

In all these flower beds Box was the favorite edging, but many other trim edgings have been used in parterres and borders by those who love not Box. Bricks were used, and boards; an edging of boards was not as pretty as one of flowers, but it kept the beds trimly in place; a garden thus edged is shown on page 63 which realizes this description of the pleasure-garden in the *Scots Gard'ner:* "The Bordures box'd and planted with variety of fine Flowers orderly Intermixt, Weeded, Mow'd, Rolled and Kept all Clean and Handsome." Germander and Rosemary were old favorites for edging. I have seen snowy edgings of Candy-tuft and Sweet Alyssum, setting off well the vari-colored blooms of the border. One of Sweet Alyssum is shown on page 256. Ageratum is a satisfactory edging. Thyme is of ancient use, but rather unmanageable; one garden owner has set his edgings of Moneywort, otherwise Creeping-jenny. I should be loth to use Moneywort as an edging; I would not care

Parterre and Clipped Box at Hampton.

for its yellow flowers in that place, though I find them very kindly and cheerful on dull banks or in damp spots, under the drip of trees and eaves, or better still, growing gladly in the flower pot of the poor. I fear if Moneywort thrived enough to make a close, suitable edging, that it would thrive too well, and would swamp the borders with its underground runners. The name Moneywort is akin to its older title Herb-twopence, or Twopenny Grass. Turner (1548) says the latter name was given from the leaves all "standying together of ech syde of the stalke lyke pence." The striped leaves of one variety of Day Lily make pretty edgings. Those from a Salem garden are here shown.

We often see in neglected gardens in New England, or by the roadside where no gardens now exist, a dense gray-green growth of Lavender Cotton, "the female plant of Southernwood," which was brought here by the colonists and here will ever remain. It was used as an edging, and is very pretty when it can be controlled. I know two or three old gardens where it is thus employed.

Sometimes in driving along a country road you are startled by a concentration of foliage and bloom, a glimpse of a tiny farm-house, over which are clustered and heaped, and round which are gathered, close enough to be within touch from door or window, flowers in a crowded profusion ample to fill a large flower bed. Such is the mass of June bloom at Wilbur Farm in old Narragansett (page 290) — a home of flowers and bees. Often by the side of the farm-house is a little garden or flower bed containing some splendid examples of old-time flowers. The splendid "running ribbons" of Snow Pinks, on page 292, are in another Narragansett garden that is a bower of blossoms. Thrift has been a common edging since the days of the old herbalist Gerarde.

"We have a bright little garden, down on a sunny slope,
Bordered with sea-pinks and sweet with the songs and blossoms of hope."

The garden of Secretary William H. Seward (in Auburn, New York), so beloved by him in his lifetime, is shown on page 146 and facing page 134. In this garden some beds are edged with Periwinkle, others with Polyanthus, and some with Ivy which Mr. Seward brought from Abbotsford in 1836. This garden was laid out in its present form in 1816, and the sun-dial was then set in its place. The garden has been enlarged, but not changed, the old "George II. Roses" and York and Lancaster Roses still grow and blossom, and the lovely arches of single Michigan Roses still flourish. In it are many flowers and fruits unusual in America, among them a bed of Alpine strawberries.

King James I. of Scotland thus wrote of the garden which he saw from his prison window in Windsor Castle: —

"A Garden fair, and in the Corners set
An Herbere greene, with Wandis long and small
Railit about."

These wandis were railings which were much used before Box edgings became universal. Sometimes they were painted the family colors, as at Hampton Court they were green and white, the Tudor colors. These "wandis" still are occasionally seen. In the Berkshire Hills I drove past an old garden thus trimly enclosed in little beds. The rails were painted a dull light brown, almost the color of some tree trunks; and Larkspur, Foxglove, and other tall flowers crowded up to them and hung their heads over the top rails as children hang over a fence or a gate. I thought it a neat, trim fashion, not one I would care for in my own garden, yet not to be despised in the garden of another.

A garden enclosed! so full of suggestion are these simple words to me, so constant is my thought that

Garden of Mrs. Mabel Osgood Wright, Waldstein, Fairfield, Conn.

an ideal flower garden must be an enclosed garden, that I look with regret upon all beautiful flower beds that are not enclosed, not shut in a frame of green hedges, or high walls, or vine-covered fences and dividing trees. It may be selfish to hide so much beauty from general view; but until our dwelling-houses are made with uncurtained glass walls, that all the world may see everything, let those who have ample grounds enclose at least a portion for the sight of friends only.

In the heart of Worcester there is a fine old mansion with ample lawns, great trees, and flowering shrubs that all may see over the garden fence as they pass by. Flowers bloom lavishly at one side of the house; and the thoughtless stroller never knows that behind the house, stretching down between the rear gardens and walls of neighboring homes, is a long enclosure of loveliness — sequestered, quiet, full of refreshment to the spirits. We think of the "Old Garden" of Margaret Deland: —

> "The Garden glows
> And 'gainst its walls the city's heart still beats.
> And out from it each summer wind that blows
> Carries some sweetness to the tired streets!"

There is a shaded walk in this garden which is a thing of solace and content to all who tread its pathway; a bit is shown opposite this page, overhung with shrubs of Lilac, Syringa, Strawberry Bush, Flowering Currant, all the old treelike things, so fair-flowered and sweet-scented in spring, so heavy-leaved and cool-shadowing in midsummer: what pleasure would there be in this shaded walk if this garden were separated from the street only by stone curbing or a low rail? And there is an old sun-dial too in this enclosed garden! I fear the street imps of a crowded city would quickly destroy the old monitor were it in an open garden; and they would make sad havoc, too, of the Roses and Larkspurs (page 65) so tenderly reared by the two sisters who

Shaded Walk in Garden of Miss Harriet P. F. Burnside, Worcester, Massachusetts.

together loved and cared for this "garden enclosed." Great trees are at the edges of this garden, and the line of tall shrubs is carried out by the lavish vines and Roses on fences and walls. Within all this

Roses and Larkspur in the Garden of Miss Harriet P. F. Burnside, Worcester, Massachusetts.

border of greenery glow the clustered gems of rare and beautiful flowers, till the whole garden seems like some rich jewel set purposely to be worn in honor over the city's heart — a clustered jewel, not one to be displayed carelessly and heedlessly.

Salem houses and gardens are like Salem people. Salem houses present to you a serene and dignified front, gracious yet reserved, not thrusting forward their choicest treasures to the eyes of passing strangers; but behind the walls of the houses, enclosed from public view, lie cherished gardens, full of the beauty of life. Such, in their kind, are Salem folk.

I know no more speaking, though silent, criticism than those old Salem gardens afford upon the modern fashion in American towns of pulling down walls and fences, removing the boundaries of lawns, and living in full view of every passer-by, in a public grassy park. It is pleasant, I suppose, for the passer-by; but homes are not made for passers-by. Old Salem gardens lie behind the house, out of sight — you have to hunt for them. They are terraced down if they stretch to the water-side; they are enclosed with hedges, and set behind high vine-covered fences, and low out-buildings; and planted around with great trees: thus they give to each family that secluded centring of family life which is the very essence and being of a home. I sat through a June afternoon in a Salem garden whose gate is within a stone's throw of a great theatre, but a few hundred feet from lines of electric cars and a busy street of trade, scarce farther from lines of active steam cars, and with a great power house for a close neighbor. Yet we were as secluded, as embowered in vines and trees, with beehives and rabbit hutches and chicken coops for happy children at the garden's end, as truly in beautiful privacy, as if in the midst of a hundred acres. Could the sense of sound be as sheltered

The Homely Back Yard.

by the enclosing walls as the sense of sight, such a garden were a city paradise.

There is scant regularity in shape in Salem gardens; there is no search for exact dimensions. Little narrow strips of flower beds run down from the main garden in any direction or at any angle where the fortunate owner can buy a few feet of land. Salem gardens do not change with the whims of fancy, either in the shape or the planting. A few new flowers find place there, such as the *Anemone Japonica* and the Japanese shrubs; for they are akin in flower sentiment, and consort well with the old inhabitants. There are many choice flowers and fruits in these gardens. In the garden of the Manning homestead (opposite page 112) grows a flourishing Fig tree, and other rare fruits; for fifty years ago this garden was known as the Pomological Garden. It is fitting it should be the home of two Robert Mannings—both well-known names in the history of horticulture in Massachusetts.

The homely back yard of an old house will often possess a trim and blooming flower border cutting off the close approach of the vegetable beds (see opposite page 66). These back yards, with the covered Grape arbors, the old pumps, and bricked paths, are cheerful, wholesome places, generally of spotless cleanliness and weedless flower beds. I know one such back yard where the pump was the first one set in the town, and children were taken there from a distance to see the wondrous sight. Why are all the old appliances for raising water so pleasing? A well-sweep is of course picturesque, with its long swinging pole, and you seem to feel the refreshment and purity of the water when you see it brought up from such a distance; and an old

Covered Well at Home of Bishop Berkeley, Whitehall, Rhode Island.

roofed well with bucket, such as this one still in use at Bishop Berkeley's Rhode Island home is ever a homelike and companionable object. But a pump is really an awkward-looking piece of mechanism, and hasn't a vestige of beauty in its lines; yet it has something satisfying about it; it may be its domesticity, its homeliness, its simplicity. We have gained infinitely in comfort in our perfect water systems and lavish water of to-day, but we have lost the gratification of the senses which came from the sight and sound of freshly drawn or running water. Much of the delight in a fountain comes, not only from the beauty of its setting and the graceful shape of its jets, but simply from the sight of the water.

Sometimes a graceful and picturesque growth of vines will beautify gate posts, a fence, or a kitchen doorway in a wonderfully artistic and pleasing fashion. On page 70 is shown the sheltered doorway of the kitchen of a fine old stone farm-house called, from its hedges of Osage Orange, "The Hedges." It stands in the village of New Hope, County Bucks, Pennsylvania. In 1718 the tract of which this farm of over two hundred acres is but a portion was deeded by the Penns to their kinsman, the direct ancestor of the present owner, John Schofield Williams, Esq. This is but one of the scores of examples I know where the same estate has been owned in one family for nearly two centuries, sometimes even for two hundred and fifty years; and in several cases where the deed from the Indian sachem to the first colonist is the only deed there has ever been, the estate having never changed ownership save by direct bequest. I have three such cases among my own kinsfolk.

Another form of garden and mode of planting which was in vogue in the "early thirties" is shown facing page 92. This pillared house and the stiff garden are excellent types; they are at Napanock, County Ulster, New York. Such a house and grounds indicated the possession of considerable wealth when they were built and laid out, for both were costly. The semicircular driveway swept up

Kitchen Doorway and Porch at the Hedges.

to the front door, dividing off Box-edged parterres like those of the day of Queen Anne. These parterres were sparsely filled, the sunnier beds being set with Spring bulbs; and there were always the yellow Day Lilies somewhere in the flower beds, and the white and blue Day Lilies, the common Funkias. Formal urns were usually found in the parterres and sometimes a great cone or ball of clipped Box. These gardens had some universal details, they always had great Snowball bushes, and Syringas, and usually white Roses, chiefly Madame Plantiers; the piazza trellises had old climbing Roses, the Queen of the Prairie or Boursault Roses. These gardens are often densely overshadowed with great evergreen trees grown from the crowded planting of seventy years ago; none are cut down, and if one dies its trunk still stands, entwined with Woodbine. I don't know that we would lay out and plant just such a garden to-day, any more than we would build exactly such a house; but I love to see both, types of the refinement of their day, and I deplore any changes. An old Southern house of allied form is shown on page 72, and its garden facing page 70,—Greenwood, in Thomasville, Georgia; but of course this garden has far more lavish and rich bloom. The decoration of this house is most interesting—a conventionalized Magnolia, and the garden is surrounded with splendid Magnolias and Crape Myrtles. The border edgings in this garden are lines of bricks set overlapping in a curious manner. They serve to keep the beds firmly in place, and the bricks are covered over with an inner edging of thrifty Violets. Curious tubs and boxes for plants are made of bricks set solidly in mortar. The garden is glorious with Roses, which seem to consort so well with Magnolias and Violets.

I love a Dutch garden, "circummured" with brick. By a Dutch garden, I mean a small garden, oblong or square, sunk about three or four feet in a lawn — so that when surrounded by brick walls they seem about two feet high when viewed outside, but are five feet or more high from within the garden. There are brick or stone steps in the middle of each of the four walls by which to descend to the garden, which may be all planted with flowers, but preferably should have set borders of flowers with

Greenwood, Thomasville, Georgia.

a grass-plot in the centre. On either side of the steps should be brick posts surmounted by Dutch pots with plants, or by balls of stone. Planted with bulbs, these gardens in their flowering time are, as old Parkinson said, a "perfect fielde of delite." We have very pretty Dutch gardens, so called, in America, but their chief claim to being Dutch is that they are set with bulbs, and have Delft or other earthen pots or boxes for formal plants or shrubs.

Sunken gardens should be laid out under the supervision of an intelligent landscape architect; and even then should have a reason for being sunken other than a whim or increase in costliness. I visited last summer a beautiful estate which had a deep sunken Dutch garden with a very low wall. It lay at the right side of the house at a little distance; and beyond it, in full view of the peristyle, extended the only squalid objects in the horizon. A garden on the level, well planted, with distant edging of shrubbery, would have hidden every ugly blemish and been a thing of beauty. As it is now, there can be seen from the house nothing of the Dutch garden but a foot or two of the tops of several clipped trees, looking like very poor, stunted shrubs. I must add that this garden, with its low wall, has been a perfect man-trap. It has been evident that often, on dark nights, workmen who have sought a "short cut" across the grounds have fallen over the shallow wall, to the gardener's sorrow, and the bulbs' destruction. Once, at dawn, the unhappy gardener found an ancient horse peacefully feeding among the Hyacinths and Tulips. He said he didn't like the grass in his new pasture nor the sudden approach to it; that he was too old for such new-fangled ways. I know another estate near Philadelphia, where the sinking of a garden revealed an exquisite view of distant hills; such a garden has reason for its form.

We have had few water-gardens in America till recent years; and there are some drawbacks to their presence near our homes, as I was vividly aware when I visited one in a friend's garden early in May this year. Water-hyacinths were even then in bloom, and two or three exquisite Lilies; and the Lotus leaves rose up charmingly from the surface of the tank. Less charmingly rose up also a cloud of vicious mosquitoes, who greeted the new-comer with a warm chorus of welcome. As our newspapers at that time were filled with plans for the application of kerosene to every inch of water-surface, such as I saw in these Lily tanks, accompanied by magnified drawings of dreadful malaria-bearing insects, I fled from them, preferring to resign both *Nymphæa* and *Anopheles*.

After the introduction to English folk of that wonder of the world, the Victoria Regia, it was cultivated by enthusiastic flower lovers in America, and was for a time the height of the floral fashion. Never has the glorious Victoria Regia and scarce any other flower been described as by Colonel Higginson, a wonderful, a triumphant word picture. I was a very little child when I saw that same lovely Lily in leaf and flower that he called his neighbor; but I have never forgotten it, nor how afraid I was of it; for some one wished to lift me upon the great leaf to see whether it would hold me above the water. We had heard that the native children in South America floated on the leaves. I objected to this experiment with vehemence; but my mother noted that I was no more frightened than was the faithful gardener at the thought of the possible strain on his precious plant of the weight of a sturdy child of six or seven years.

Roses and Violets in Garden at Greenwood, Thomasville, Georgia.

Water Garden at Sylvester Manor, Shelter Island, New York.

I have seen the Victoria Regia leaves of late years, but I seldom hear of its blossoming; but alas! we take less heed of the blooming of unusual plants than we used to thirty or forty years ago. Then people thronged a greenhouse to see a new Rose or Camellia Japonica; even a Night-blooming Cereus attracted scores of visitors to any house where it blossomed. And a fine Cactus of one of our neighbors always held a crowded reception when in rich bloom. It was a part of the "Flower Exchange," an interest all had for the beautiful flowers of others, a part of the old neighborly life.

Within the past five or six years there have been laid out in America, at the country seats of men of wealth and culture, a great number of formal gardens, — Italian gardens, some of them are worthily named, as they have been shaped and planted in conformity with the best laws and rules of Italian garden-making — that special art. On this page is shown the finely proportioned terrace wall, and opposite the upper terrace and formal garden of Drumthwacket, Princeton, New Jersey, the country seat of M. Taylor Pyne, Esq. This garden affords a good example of the accord which should ever exist

Terrace Wall at Drumthwacket, Princeton, New Jersey.

between the garden and its surroundings. The name, Drumthwacket — a wooded hill — is a most felicitous one; the place is part of the original grant to William Penn, and has remained in the possession of one family until late in the nineteenth century. From this beautifully wooded hill the terrace-garden overlooks the farm buildings, the linked ponds, the fertile fields and meadows; a serene pastoral view, typical of the peaceful landscape of that vicinity —

Garden at Avonwood Court, Haverford, Pennsylvania, Country-seat of Charles E. Mather, Esq.

yet it was once the scene of fiercest battle. For the Drumthwacket farm is the battle-ground of that important encounter of 1777 between the British and the Continental troops, known as the Battle of Princeton, the turning point of the Revolution, in which Washington was victorious. To this day,

Garden at Drumthwacket, Princeton, New Jersey.

cannon ball and grape shot are dug up in the Drumthwacket fields. The Lodge built in 1696 was, at Washington's request, the shelter for the wounded British officers; and the Washington Spring in front of the Lodge furnished water to Washington. The group of trees on the left of the upper pond marks the sheltered and honored graves of the British soldiers, where have slept for one hundred and twenty-four years those killed at this memorable encounter. If anything could cement still more closely the affections of the English and American peoples, it would be the sight of the tenderly sheltered graves of British soldiers in America, such as these at Drumthwacket and other historic fields on our Eastern coast. At Concord how faithfully stand the sentinel pines over the British dead of the Battle of Concord, who thus repose, shut out from the tread of heedless feet yet ever present for the care and thought of Concord people.

We have older Italian gardens. Some of them are of great loveliness, among them the unique and dignified garden of Hollis H. Hunnewell, Esq., but many of the newer ones, even in their few summers, have become of surprising grace and beauty, and their exquisite promise causes a glow of delight to every garden lover. I have often tried to analyze and account for the great charm of a formal garden, to one who loves so well the unrestrained and lavished blossoming of a flower border crowded with nature-arranged and disarranged blooms. A chance sentence in the letter of a flower-loving friend, one whose refined taste is an inherent portion of her nature, runs thus: —

> "I have the same love, the same sense of perfect satisfaction, in the old formal garden that I have in the sonnet in poetry, in the Greek drama as contrasted with the modern drama; something within me is ever drawn toward that which is restrained and classic."

In these few words, then, is defined the charm of the formal garden — a well-ordered, a classic restraint.

Some of the new formal gardens seem imperfect in design and inadequate in execution; worse still, they are unsuited to their surroundings; but gracious nature will give even to these many charms of color, fragrance, and shape through lavish plant growth. I have had given to me sets of beautiful photographs of these new Italian gardens, which I long to include with my pictures of older flower beds; but I cannot do so in full in a book on Old-time Gardens, though they are copied from far older gardens than our American ones. I give throughout my book occasional glimpses of detail in modern formal gardens; and two examples may be fitly illustrated and described in comparative fulness in this book, because they are not only unusual in their beauty and promise, but because they have in plan and execution some bearing on my special presentation of gardens. These two are the gardens of Avonwood Court in Haverford, Pennsylvania, the country-seat of Charles E. Mather, Esq., of Philadelphia; and of Yaddo, in Saratoga, New York, the country-seat of Spencer Trask, Esq., of New York.

The garden at Avonwood Court was designed and laid out in 1896 by Mr. Percy Ash. The flower planting was done by Mr. John Cope; and the garden is delightsome in proportions, contour, and aspect. Its claim to illustrative description in this book lies in the fact that it is planted chiefly with old-fashioned flowers, and its beds are laid out and bordered with thriving Box in a truly old-time mode. It affords a striking example of the beauty and satisfaction that can come from the use of Box as an edging, and old-time flowers as a filling of these beds. Among the two hundred different plants are great rows of yellow Day Lilies shown in the view facing page 76; regular plantings of Peonies; borders of Flower de Luce; banks of Lilies of the Valley; rows of white Fraxinella and Lupine, beds of fringed Poppies, sentinels of Yucca — scores of old favorites have grown and thriven in the cheery manner they ever display when they are welcome and beloved. The sun-dial in this garden is shown facing page 82; it was designed by Mr. Percy Ash, and can be regarded as a model of simple outlines, good proportions, careful placing, and symmetrical setting. By placing I mean that it is in the right site in relation to the surrounding flower beds, and to the general outlines of the garden; it is a dignified and significant garden centre. By setting I mean its being raised to proper prominence in the garden scheme, by being placed at the top of a platform formed of three circular steps of ample proportion and suitable height, that its pedestal is also of the right size and not so high but one can, when standing on the top step, read with ease the dial's response to our question, "What's the time o' the day?" The hedges and walls of Honeysuckle, Roses, and other flowering vines that surround this garden have thriven wonderfully in the five years of the garden's life, and look like settings of many years. The simple but graceful wall seat gives

Sundial at Avonwood Court, Haverford, Pennsylvania.

Entrance Porch and Gate to the Rose Garden at Yaddo.

some idea of the symmetrical and simple garden furnishings, as well as the profusion of climbing vines that form the garden's boundaries.

This book bears on the title-page a redrawing of a charming old woodcut of the eighteenth century, a very good example of the art thought and art execution of that day, being the work of a skilful designer. It is from an old stilted treatise on orchards and gardens, and it depicts a cheerful little Love, with anxious face and painstaking care, measuring and laying out the surface of the earth in a garden. On his either side are old clipped Yews; and at his feet a spade and pots of garden flowers, among them the Fritillary so beloved of all flower lovers and herbalists of that day, a significant flower — a flower of meaning and mystery. This drawing may be taken as an old-time emblem, and a happy one, to symbolize the making of the beautiful modern Rose Garden at Yaddo; where Love, with tenderest thought, has laid out the face of the earth in an exquisite garden of Roses, for the happiness and recreation of a dearly loved wife. The noble entrance gate and porch of this Rose Garden formed a happy surprise to the garden's mistress when unveiled at the dedication of the garden. They are depicted on page 81, and there may be read the inscription which tells in a few well-chosen words the story of the inspiration of the garden; but "between the lines," to those who know the Rose Garden and its makers, the inscription speaks with even deeper meaning the story of a home whose beauty is only equalled by the garden's spirit. To all such readers the Rose Garden becomes a fitting expression of the life of those who own it and care for it. This quality of expression, of significance, may be seen in many a smaller and simpler garden, even in a tiny cottage plot; you can perceive, through the care bestowed upon it, and its responsive blossoming, a *something* which shows the life of the garden owners; you know that they are thoughtful, kindly, beauty-loving, home-loving.

Behind the beautiful pergola of the Yaddo garden, set thickly with Crimson Rambler, a screenlike row of poplars divides the Rose Garden from a luxuriant Rock Garden, and an Old-fashioned Garden of large extent, extraordinary profusion, and many years' growth. Perhaps the latter-named garden might seem more suited to my pages, since it is more advanced in growth and apparently more akin to my subject; but I wish to write specially of the Rose Garden, because it is an unusual example of what can be accomplished without aid of architect or landscape gardener, when good taste, care-

Pergola and Terrace Walk in Rose Garden at Yaddo.

ful thought, attention to detail, a love of flowers, and *intent to attain perfection* guide the garden's makers. It is happily placed in a country of most charming topography, but it must not be thought that the garden shaped itself; its beautiful proportions, contour, and shape were carefully studied out and brought to the present perfection by the same force that is felt in the garden's smallest detail, the power of Love. The Rose Garden is unusually large for a formal garden; with its vistas and walks, the connected Daffodil Dell, and the Rock Garden, it fills about ten acres. But the

Statue of Christalan in Rose Garden at Yaddo.

estate is over eight hundred acres, and the house very large in ground extent, so the garden seems well-proportioned. This Rose Garden has an unusual attraction in the personal interest of every detail, such as is found in few American gardens of great size, and indeed in few English gardens. The gardens of the Countess Warwick, at Easton Lodge, in Essex, possess the same charm, a personal meaning and significance in the statues and fountains, and even in the planting of flower borders. The illustration on page 83 depicts the general shape of the Yaddo Rose Garden, as seen from the upper terrace; but it does not show how the garden stretches down the fine marble steps, past the marble figures of Diana and Paris, and along the paths of standard Roses, past the shallow fountain which is not so large as to obscure what speaks the garden's story, the statue of Christalan, that grand creation in one of Mrs. Trask's idyls, *Under King Constantine.* This heroic figure, showing to full extent the genius of the sculptor, William Ordway Partridge, also figures the genius of the poet-creator, and is of an inexpressible and impressive nobility. With hand and arm held to heaven, Christalan shows against the background of rich evergreens as the true knight of this garden of sentiment and chivalry.

> "The sunlight slanting westward through the trees
> Fell first upon his lifted, golden head,
> Making a shining helmet of his curls,
> And then upon the Lilies in his hand.
> His eyes had a defiant, fearless glow;
> Against the sombre background of the wood
> He looked scarce human."

The larger and more impressive fountain at Yaddo is shown on these pages. It is one hundred feet long and seventy feet wide, and is in front of the house, to the east. Its marble figures signify the Dawn;

it will be noted that on this site its beauties show against a suited and ample background, and its

Sun-dial in Rose Garden at Yaddo.

grand proportions are not permitted to obscure the fine statue of Christalan from the view of those seated on the terrace or walking under the shade of the pergola.

Especially beautiful is the sun-dial on the upper terrace, shown on page 86. The metal dial face is supported by a marble slab resting on two carved standards of classic design representing conventionalized lions, these being copies of those two splendid standards unearthed at Pompeii, which still may be seen by the side of the impluvium in the atrium or main hall of the finest Græco-Roman dwelling-place which has been restored in that wonderful city. These sun-dial standards at Yaddo were made by the permission and under the supervision of the Italian government. I can conceive nothing more fitting or more inspiring to the imagination than that, telling as they do the story of the splendor of ancient Pompeii and of the passing centuries, they should now uphold to our sight a sun-dial as if to bid us note the flight of time and the vastness of the past.

Bronze Face of Dial in Rose Garden at Yaddo.

The entire sun-dial, with its beautiful adjuncts of carefully shaped marble seats, stands on a semicircular plaza of marble at the head of the noble flight of marble steps. The engraved metal dial face bears two exquisite verses — the gift of one poet to another — of Dr. Henry Van Dyke to the garden's mistress, Katrina Trask. These dial mottoes are unusual, and perfect examples of that genius which with a few words can shape a lasting gem of our English tongue. At the edge of the dial face is this motto:

"Hours fly,
Flowers die,
New Days,
New Ways,
Pass by;
Love stays."

At the base of the gnomon is the second motto: —

Time is
Too Slow for those who Wait,
Too Swift for those who Fear,
Too Long for those who Grieve,
Too Short for those who Rejoice;
But for those who Love,
Time is
Eternity.

I have for years been a student of sun-dial lore, a collector of sun-dial mottoes and inscriptions, of which I have many hundreds. I know nowhere, either in English, on English or Scotch sun-dials, or in the Continental tongues, any such exquisite dial legends as these two — so slight of form, so simple in wording, so pure in diction, yet of sentiment, of thought, how full! how impressive! They stamp themselves forever on the memory as beautiful examples of what James Russell Lowell called verbal magic; that wonderful quality which comes, neither from chosen words, nor from their careful combination into sentences, but from something

Ancient Pine in Garden at Yaddo.

which is as inexplicable in its nature as it is in its charm.

To tree lovers the gardens and grounds at Yaddo have glorious charms in their splendid trees; but one can be depicted here — the grand native Pine, over eight feet in diameter, which, with other stately sentinels of its race, stands a sombrely beautiful guard over all this loveliness.

CHAPTER IV

BOX EDGINGS

"They walked over the crackling leaves in the garden, between the lines of Box, breathing its fragrance of eternity; for this is one of the odors which carry us out of time into the abysses of the unbeginning past; if we ever lived on another ball of stone than this, it must be that there was Box growing on it."

— *Elsie Venner*, Oliver Wendell Holmes, 1861.

O many of us, besides Dr. Holmes, the unique aroma of the Box, cleanly bitter in scent as in taste, is redolent of the eternal past; it is almost hypnotic in its effect. This strange power is not felt by all, nor is it a present sensitory influence; it is an hereditary memory, half-known by many, but fixed in its intensity in those of New England birth and descent, true children of the Puritans; to such ones the Box breathes out the very atmosphere of New England's past. I cannot see in clear outline those prim gardens of centuries ago, nor the faces of those who walked and worked therein; but I know, as I stroll to-day between our old Box-edged borders, and inhale the beloved bitterness of fragrance, and gather a stiff sprig of the beautiful glossy leaves, that in truth the garden lovers and garden workers of other days walk beside me, though unseen and unheard.

About thirty years ago a bright young Yankee girl went to the island of Cuba as a governess to the family of a sugar planter. It was regarded as a somewhat perilous adventure by her home-staying folk, and their apprehensions of ill were realized in her death there five years later. This was not, however, all that happened to her. The planter's wife had died in this interval of time, and she had been married to the widower. A daughter had been born, who, after her mother's death, was reared in the Southern island, in Cuban ways, having scant and formal communication with her New England kin. When this girl was twenty years old, she came to the little Massachusetts town where her mother had been reared, and met there a group of widowed and maiden aunts, and great-aunts. After sitting for a time in her mother's room in the old home, the reserve which often exists between those of the same race who should be friends but whose lives have been widely apart, and who can never have more than a passing sight of each other, made them in semi-embarrassment and lack of resources of mutual interest walk out into the garden. As they passed down the path between high lines of Box, the girl suddenly stopped, looked in terror at the gate, and screamed out in fright, "The dog, the dog, save me, he will kill me!" *No dog was there*, but on that very spot, between those Box hedges, thirty years before, her mother had been attacked and bitten by an enraged dog, to the distress and apprehension of

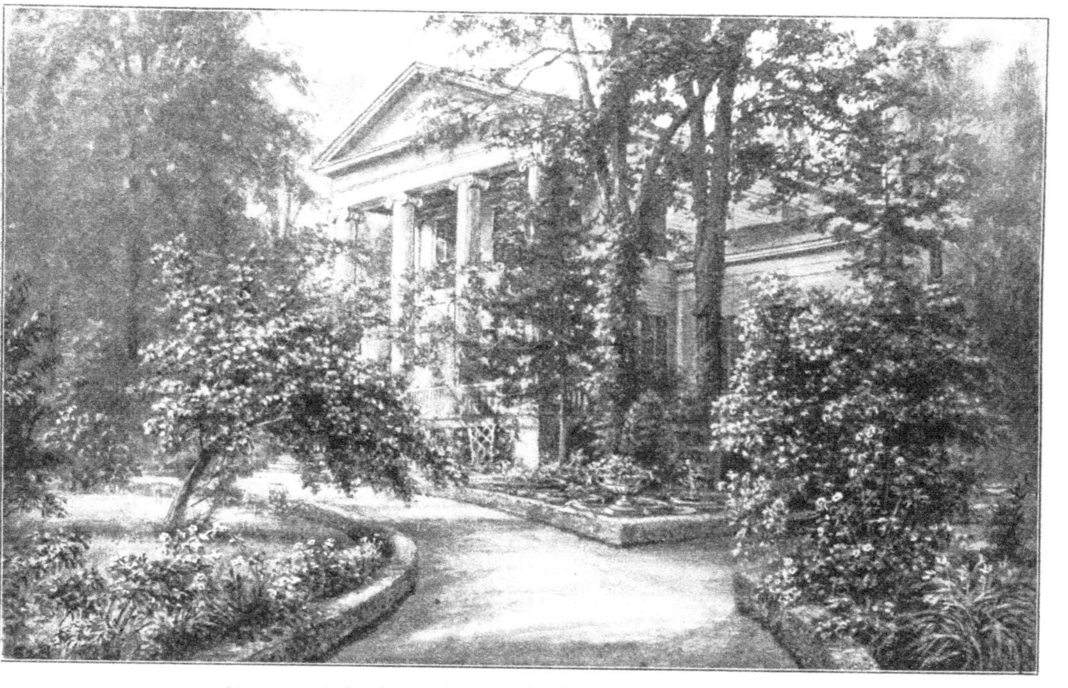

House and Garden at Napanock, County Ulster, New York.

the aunts, who all recalled the occurrence, as they reassured the fainting and bewildered girl. She, of course, had never known aught of this till she was told it by the old Box.

Many other instances of the hypnotic effect of Box are known, and also of its strong influence on the mind through memory. I know of a man who travelled a thousand miles to renew acquaintance and propose marriage to an old sweetheart, whom he had not seen and scarcely thought of for years, having been induced to this act wholly through memories of her, awakened by a chance stroll in an old Box-edged garden such as those of his youth; at the gate of one of which he had often lingered, after walking home with her from singing-school. I ought to be able to add that the twain were married as a result of this sentimental memory-awakening through the old Box; but, in truth, they never came very close to matrimony. For when he saw her he remained absolutely silent on the subject of marriage; the fickle creature forgot the Box scent and the singing-school, while she openly expressed to her friends her surprise at his aged appearance, and her pity for his dulness. For the sense of sight is more powerful than that of smell, and the Box might prove a master hand at hinting, but it failed utterly in permanent influence.

Those who have not loved the Box for centuries in the persons and with the partial noses of their Puritan forbears, complain of its curious scent, say, like Polly Peacham, that "they can't abear it," and declare that it brings ever the thought of old graveyards. I have never seen Box in ancient burying-grounds, they were usually too neglected to be thus planted; but it was given a limited space in the cemeteries of the middle of this century. Even those borders have now generally been dug up to give place to granite copings.

The scent of Box has been aptly worded by Gabriel d'Annunzio, in his *Virgin of the Rocks*, in his description of a neglected garden. He calls it a "bitter sweet odor," and he notes its influence in making his wanderers in this garden "reconstruct some memory of their far-off childhood."

The old Jesuit poet Rapin writing in the seventeenth century tells a fanciful tale that—

> "Gardens of old, nor Art, nor Rules obey'd,
> But unadorn'd, or wild Neglect betray'd;"

that Flora's hair hung undressed, neglected "in artless tresses," until in pity another nymph "around her head wreath'd a Boxen Bough" from the fields; which so improved her beauty that trim edgings were placed ever after—"where flowers disordered once at random grew."

He then describes the various figures of Box, the way to plant it, its disadvantages, and the associate flowers that should be set with it, all in stilted verse.

Queen Anne was a royal enemy of Box. By her order many of the famous Box hedges at Hampton Court were destroyed; by her example, many old Box-edged gardens throughout England were rooted up. There are manifold objections raised to Box besides the dislike of its distinctive odor: heavy edgings and hedges of Box "take away the heart of the ground" and flowers pine within Box-edged borders; the roots of Box on the inside of the

Box Parterre at Hampton.

flower knot or bed, therefore, have to be cut and pulled out in order to leave the earth free for flower roots. It is also alleged that Box harbors slugs—and I fear it does.

We are told that it is not well to plant Box edgings in our gardens, because Box is so frail, is so easily winter-killed, that it dies down in ugly fashion. Yet see what great trees it forms, even when untrimmed, as in the Prince Garden (page 31). It is true that Box does not always flourish in the precise shape you wish, but it has nevertheless a wonderfully tenacious hold on life. I know nothing more suggestive of persistence and of sad sentiment than the view often seen in forlorn city enclosures, as you drive past, or rush by in an electric car, of an aged bush of Box, or a few feet of old Box hedge growing in the beaten earth of a squalid back yard, surrounded by dirty tenement houses. Once a fair garden there grew; the turf and flowers and trees are vanished; but spared through accident, or because deemed so valueless, the Box still lives. Even in Washington and other Southern cities, where the negro population eagerly gather Box at Christmastide, you will see these forlorn relics of the garden still growing, and their bitter fragrance rises above the vile odors of the crowded slums.

Box formed an important feature of the garden of Pliny's favorite villa in Tuscany, which he described in his letter to Apollinaris. How I should have loved its formal beauty! On the southern front a terrace was bordered with a Box hedge and "embellished with various figures in Box, the representation of divers animals." Beyond was a circus formed around by ranges of Box rising in walls of varied heights. The middle of this circus was ornamented with figures of Box. On one side was a hippodrome set with a plantation of Box trees backed with Plane trees; thence ran a straight walk divided by Box hedges into alleys. Thus expanses were enclosed, one of which held a beautiful meadow, another had "knots of Plane tree," another was "set with Box a thousand different forms." Some of these were letters expressing the name of the owner of all this extravagance; or the initials of various fair Roman dames, a very gallant pleasantry of young Pliny. Both Plane tree and Box tree of such ancient gardens were by tradition nourished with wine instead of water. Initials of Box may be seen to-day in English gardens, and heraldic devices. French gardens vied with English gardens in curious patterns in Box. The garden of Versailles during the reign of Louis XIV. had a stag chase, in clipped Box, with greyhounds in chase. Globes, pyramids, tubes, cylinders, cones, arches, and other shapes were cut in Box as they were in Yew.

A very pretty conceit in Box was —

> "Horizontal dials on the ground
> In living Box by cunning artists traced."

Reference is frequent enough to these dials of Box to show that they were not uncommon in fine old English gardens. There were sun-dials either of Box or Thrift, in the gardens of colleges both at Oxford and Cambridge, as may be seen in Loggan's *Views*. Two modern ones are shown; one, on page 98, is in the garden of Lady Lennox, at Broughton Castle, Banbury, England. Another of exceptionally fine growth and trim perfection in the garden at Ascott, the seat of Mr. Leopold de Rothschild (opposite page 100.) These are curious rather than beautiful, but display well that quality given in the poet's term "the tonsile Box."

Sun-dial in Box at Broughton Castle.

Writing of a similar sun-dial, Lady Warwick says:—

"Never was such a perfect timekeeper as my sun-dial, and the figures which record the hours are all cut out and trimmed in Box, and there again on its outer ring is a legend which read in whatever way you please: Les heures heureuses ne se comptent pas. They were outlined for me, those words, in baby sprigs of Box by a friend who is no more, who loved my garden and was good to it."

Box hedges were much esteemed in England — so says Parkinson, to dry linen on, affording the raised expanse and even surface so much desired. It can always be noted in all domestic records of early days that the vast washing of linen and clothing was one of the great events of the year. Sometimes, in households of plentiful supply, these washings were done but once a year; in other homes, semi-annually. The drying and bleaching linen was an unceasing attraction to rascals like Autolycus, who had a "pugging tooth" — that is, a prigging tooth. These linen thieves had a special name, they were called "prygmen"; they wandered through the country on various pretexts, men and their doxies, and were the bane of English housewives.

The Box hedges were also in constant use to hold the bleaching webs of homespun and woven flaxen and hempen stuff, which were often exposed for weeks in the dew and sunlight. In 1710 a reason given for the disuse and destruction of "quicksetted arbors and hedges" was that they "agreed very ill with the ladies' muslins."

Box was of little value in the apothecary shop, was seldom used in medicine. Parkinson said that the leaves and dust of boxwood "boyld in lye" would make hair to be "of an Aborne or Abraham color" — that is, auburn. This was a very primitive hair dye, but it must have been a powerful one.

Boxwood was a firm, beautiful wood, used to make tablets for inscriptions of note. The mottled wood near the root was called dudgeon. Holland's translation of Pliny says, "The Box tree seldome hath any grain crisped damaske-wise, and never but about the root, the which is dudgin." From its esteemed use for dagger hilts came the word dudgeon-dagger, and the terms "drawn-dudgeon" and "high-dudgeon," meaning offence or discord.

I plead for the Box, not for its fragrance, for you may not be so fortunate as to have a Puritan sense of smell, nor for its weird influence, for that is intangible; but because it is the most becoming of all edgings to our garden borders of old-time flowers. The clear compact green of its shining leaves, the trim distinctness of its clipped lines, the attributes that made Pope term it the "shapely Box," make it the best of all foils for the varied tints of foliage, the many colors of bloom, and the careless grace in growth of the flowers within the border.

Box edgings are pleasant, too, in winter, showing in grateful relief against the tiresome monotony of the snow expanse. And they bear sometimes a crown of lightest snow wreaths, which seem like a white blossoming in promise of the beauties of the border in the coming summer. Pick a bit of this winter Box, even with the mercury below zero. Lo! you have a breath of the hot dryness of the midsummer garden.

Box grows to great size, even twenty feet in height. In Southern gardens, where it is seldom winter-killed, it is often of noble proportions. In the lovely garden of Martha Washington at Mount Vernon the Box is still preserved in the beauty and interest of its original form.

The Box edgings and hedges of many other

Sun-dial in Box at Ascott.

Southern gardens still are in good condition; those of the old Preston homestead at Columbia, South Carolina (shown on pages 15 and 18, and facing page 54), owe their preservation during the Civil War to the fact that the house was then the refuge of a sisterhood of nuns. The Ridgely estate, Hampton, in County Baltimore, Maryland, has a formal garden in which the perfection of the Box is a delight. The will of Captain Charles Ridgely, in 1787, made an appropriation of money and land for this garden. The high terrace which overlooks the garden and the shallow ones which break the southern slope and mark the boundaries of each parterre are fine examples of landscape art, and are said to be the work of Major Chase Barney, a famous military engineer. By 1829 the garden was an object of beauty and much renown. A part only of the original parterre remains, but the more modern flower borders, through the unusual perspective and contour of the garden, do not clash with the old Box-edged beds. These edgings were reset in 1870, and are always kept very closely cut. The circular domes of clipped box arise from stems at least a hundred years old. The design of the parterre is so satisfactory that I give three views of it in order to show it fully. (See pages 57, 60, and 95.)

A Box-edged garden of much beauty and large extent existed for some years in the grounds connected with the County Jail in Fitchburg, Massachusetts. It was laid out by the wife of the warden, aided by the manual labor of convicted prisoners, with her earnest hope that working among flowers would have a benefiting and softening influence on these criminals. She writes rather dubiously: "They all enjoyed being out of doors with their pipes, whether among the flowers or the vegetables; and no attempt at escape was ever made by any of them while in the comparative freedom of the flower-garden." She planted and marked distinctly in this garden over seven hundred groups of annuals and hardy perennials, hoping the men would care to learn the names of the flowers, and through that knowledge, and their practise in the care of Box edgings and hedges, be able to obtain positions as under-gardeners when their terms of imprisonment expired.

The garden at Tudor Place, the home of Mrs. Beverley Kennon (page 103), displays fine Box; and the garden of the poet Longfellow which is said to have been laid out after the Box-edged parterres at Versailles. Throughout this book are scattered several good examples of Box from Salem and other towns; in a sweet, old garden on Kingston Hill, Rhode Island (page 104) the flower-beds are anchor-shaped.

In favorable climates Box edgings may grow in such vigor as to entirely fill the garden beds. An example of this is given on page 105, showing the garden at Tuckahoe. The beds were laid out over a large space of ground in a beautiful design, which still may be faintly seen by examining the dark expanse beside the house, which is now almost solid Box. The great hedges by the avenue are also Box; between similar ones at Uhpton Court in

Camden, South Carolina, riders on horseback cannot be seen nor see over it. New England towns seldom show such growth of Box; but in Hingham, Massachusetts, at the home of Mrs. Robbins, author of that charming book, *The Rescue of an Old Place*,

Garden at Tudor Place.

there is a Box bower, with walls of Box fifteen feet in height. These walls were originally the edgings of a flower bed on the "Old Place." Read Dr. John Brown's charming account of the Box bower of the "Queen's Maries."

Box grows on Long Island with great vigor. At Brecknock Hall, the family residence of Mrs. Albert

Delafield at Greenport, Long Island, the hedges of plain and variegated Box are unusually fine, and the paths are well laid out. Some of them are entirely covered by the closing together of the two hedges which are often six or seven feet in height.

In spite of the constant assertion of the winter-killing of Box in the North, the oldest Box in

Anchor-shaped Flower-beds, Kingston, Rhode Island.

the country is that at Sylvester Manor, Shelter Island, New York. The estate is now owned by the tenth mistress of the manor, Miss Cornelia Horsford; the first mistress of the manor, Grissel Sylvester, who had been Grissel Gardiner, came there in 1652. It is told, and is doubtless true, that she brought there the first Box plants, to make, in what was then a far-away island, a semblance of her home garden. It is said that this Box was thriving in Madam Sylvester's garden when George Fox preached there to the Indians. The oldest Box is fifteen or eighteen feet high; not so tall, I think, as the neglected Box at Vaucluse, the old Hazard place near Newport, but far more massive and thrifty and shapely. Box needs unusual care and judgment, an instinct almost, for the removal of certain portions.

Ancient Box at Tuckakoe.

It sends out tiny rootlets at the joints of the sprays, and these grow readily. The largest and oldest Box bushes at Sylvester Manor garden are a study in their strong, hearty stems, their perfect foliage, their symmetry; they show their care of centuries.

The delightful Box-edged flower beds were laid out in their present form about seventy years ago by the grandfather of the present owner. There is a Lower Garden, a Terrace Garden, which are shown on succeeding pages, a Fountain Garden, a

Rose Garden, a Water Garden; a bit of the latter is on page 75. In some portions of these gardens, especially on the upper terrace, the Box is so high, and set in such quaint and rambling figures, that it closely approaches an old English maze; and it was a pretty sight to behold a group of happy little children running in and out among these Box hedges that extended high over their heads, searching long and eagerly for the central bower where their little tea party was set.

Over these old garden borders hangs literally an atmosphere of the past; the bitter perfume stimulates the imagination as we walk by the side of these splendid Box bushes, and think, as every one must, of what they have seen, of what they know; on this garden is written the history of over two centuries of beautiful domestic home life. It is well that we still have such memorials to teach us the nobility and beauty of such a life.

CHAPTER V

THE HERB GARDEN

"To have nothing here but Sweet Herbs, and those only choice ones too, and every kind its bed by itself."

— Desiderius Erasmus, 1500.

N Montaigne's time it was the custom to dedicate special chapters of books to special persons. Were it so to-day, I should dedicate this chapter to the memory of a friend who has been constantly in my mind while writing it; for she formed in her beautiful garden, near our modern city, Chicago, the only perfect herb garden I know, — a garden that is the counterpart of the garden of Erasmus, made four centuries ago; for in it are "nothing but Sweet Herbs, and choice ones too, and every kind its bed by itself." A corner of it is shown on page 108. This herb garden is so well laid out that I will give directions therefrom for a bed of similar planting. It may be placed at the base of a grass bank or at the edge of a garden. Let two garden walks be laid out, one at the lower edge, perhaps, of the bank, the other parallel, ten, fifteen, twenty feet away. Let narrow paths be left at regular intervals running parallel from walk to walk, as do the rounds of a ladder from the two side bars. In the narrow oblong beds formed by these paths plant solid rows of herbs, each variety by itself, with no attempt at diversity of design. You can thus walk among them, and into them, and smell them in their concentrated strength, and you can gather them at ease. On the bank can be placed the creeping Thyme, and other

Herb Garden at White Birches, Elmhurst, Illinois.

low-running herbs. Medicinal shrubs should be the companions of the herbs; plant these as you will, according to their growth and habit, making them give variety of outline to the herb garden.

There are few persons who have a strong enough love of leaf scents, or interest in herbs, to make them willing to spend much time in working in an herb garden. The beauty and color of flowers would compensate them, but not the growth or scent of leafage. It is impossible to describe to one who does not feel by instinct "the lure of green things growing," the curious stimulation, the sense of intoxication, of delight, brought by working among such green-growing, sweet-scented things. The maker of this interesting garden felt this stimulation and delight; and at her city home on a bleak day in December we both revelled in holding and breathing in the scent of tiny sprays of Rue, Rosemary, and Balm which, still green, had been gathered from beneath fallen leaves and stalks in her country garden, as a tender and grateful attention of one herb lover to another. Thus did she prove Shakespeare's words true even on the shores of Lake Michigan: —

> "Rosemary and Rue: these keep
> Seeming and savor all the winter long."

There is ample sentiment in the homely inhabitants of the herb garden. The herb garden of the Countess of Warwick is called by her a Garden of Sentiment. Each plant is labelled with a pottery marker, swallow-shaped, bearing in ineradicable colors the flower name and its significance. Thus there is Balm for sympathy, Bay for glory, Foxglove for sincerity, Basil for hatred.

A recent number of *The Garden* deplored the dying out of herbs in old English gardens; so I think it may prove of interest to give the list of herbs and medicinal shrubs and trees which grew in this friend's herb garden in the new world across the sea.

Arnica, Anise, Ambrosia, Agrimony, Aconite.

Belladonna, Black Alder, Betony, Boneset or Thoroughwort, Sweet Basil, Bryony, Borage, Burnet, Butternut, Balm, *Melissa officinalis*, Balm (variegated), Bee-balm, or Oswego tea, mild, false, and true Bergamot, Burdock, Bloodroot, Black Cohosh, Barberry, Bittersweet, Butterfly-weed, Birch, Blackberry, Button-Snakeroot, Buttercup.

Costmary, or Sweet Mary, Calamint, Choke-cherry, Comfrey, Coriander, Cumin, Catnip, Caraway, Chives, Castor-oil Bean, Colchicum, Cedronella, Camomile, Chicory, Cardinal-flower, Celandine, Cotton, Cranesbill, Cow-parsnip, High-bush Cranberry.

Dogwood, Dutchman's-pipe, Dill, Dandelion, Dock, Dogbane.

Elder, Elecampane, Slippery Elm.

Sweet Fern, Fraxinella, Fennel, Flax, Fumitory, Fig, Sweet Flag, Blue Flag, Foxglove.

Goldthread, Gentian, Goldenrod.

Hellebore, Henbane, Hops, Horehound, Hyssop, Horse-radish, Horse-chestnut, Hemlock, Small Hemlock or Fool's Parsley.

American Ipecac, Indian Hemp, Poison Ivy, wild, false, and blue Indigo, wild yellow Indigo, wild white Indigo.

Juniper, Joepye-weed.

Lobelia, Lovage, Lavender Lemon Verbena, Lemon, Mountain Laurel, Yellow Lady's-slippers, Lily of the Valley, Liverwort, Wild Lettuce, Field Larkspur, Lungwort.

Mosquito plant, Wild Mint, Motherwort, Mullein, Sweet Marjoram, Meadowsweet, Marshmallow, Mandrake, Mulberry, black and white Mustard, Mayweed, Mugwort, Marigold.

Nigella.

Opium Poppy, Orange, Oak.

Pulsatilla, Pellitory or Pyrethrum, Red Pepper, Pepper-

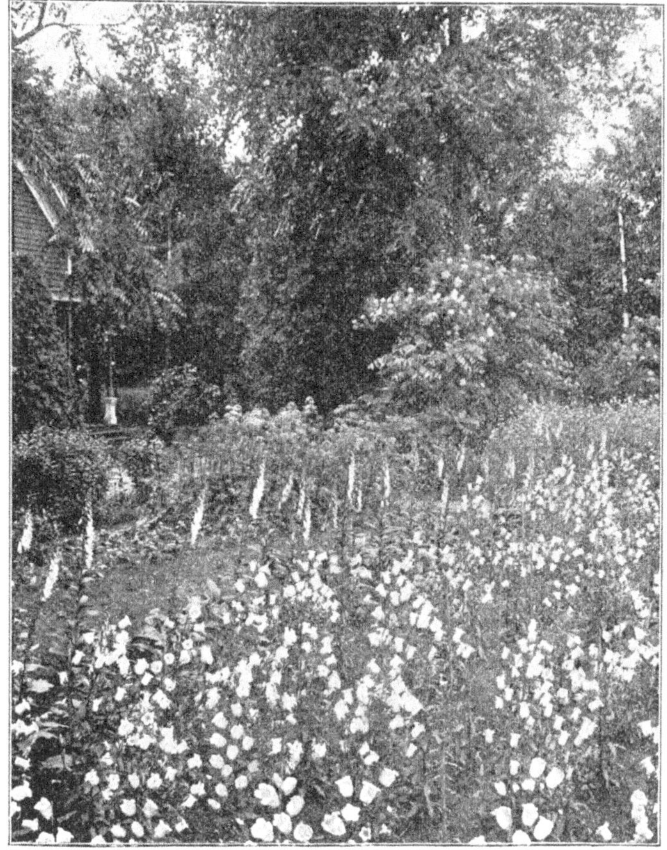

Garden at White Birches, Elmhurst, Illinois.

mint, Pennyroyal, False Pennyroyal, Pope-weed, Pine, Pigweed, Pumpkin, Parsley, Prince's-pine, Peony, Plantain.

Rhubarb, Rue, Rosemary, Rosa gallica, Dog Rose.

Sassafras, Saxifrage, Sweet Cicely, Sage (common blue), Sage (red), Summer Savory, Winter Savory, Santonin, Sweet Woodruff, Saffron, Spearmint, wild Sarsaparilla, Black Snakeroot, Squills, Senna, St.-John's-Wort, Sorrel, Spruce Fir, Self-heal, Southernwood.

Thorn Apple, Tansy, Thyme, Tobacco, Tarragon.

Valerian, Dogtooth Violet, Blue Violet.

Witchhazel, Wormwood, Wintergreen, Willow, Walnut.

Yarrow.

It will be noted that some common herbs and medicinal plants are missing; there is, for instance, no Box; it will not live in that climate; and there are many other herbs which this garden held for a short time, but which succumbed under the fierce winter winds from Lake Michigan.

It is interesting to compare this list with one made in rhyme three centuries ago, the garland of herbs of the nymph Lelipa in Drayton's *Muse's Elyzium*.

"A chaplet then of Herbs I'll make
  Than which though yours be braver,
Yet this of mine I'll undertake
  Shall not be short in savour.
With Basil then I will begin,
  Whose scent is wondrous pleasing:
This Eglantine I'll next put in
  The sense with sweetness seizing.
Then in my Lavender I lay
  Muscado put among it,
With here and there a leaf of Bay,
  Which still shall run along it.
Germander, Marjoram and Thyme,
  Which uséd are for strewing;
With Hyssop as an herb most prime

Garden of Manning Homestead, Salem, Massachusetts.

Here in my wreath bestowing.
Then Balm and Mint help to make up
My chaplet, and for trial
Costmary that so likes the Cup,
And next it Pennyroyal.
Then Burnet shall bear up with this,
Whose leaf I greatly fancy;
Some Camomile doth not amiss
With Savory and some Tansy.
Then here and there I'll put a sprig
Of Rosemary into it,
Thus not too Little nor too Big,
'Tis done if I can do it."

Another name for the herb garden was the olitory; and the word herber, or herbar, would at first sight appear to be an herbarium, an herb garden; it was really an arbor. I have such satisfaction in herb gardens, and in the herbs themselves, and in all their uses, all their lore, that I am confirmed in my belief that I really care far less for Botany than for that old-time regard and study of plants covered by the significant name, Wort-cunning. Wort was a good old common English word, lost now in our use, save as the terminal syllable of certain plant-names; it is a pity we have given it up since its equivalent, herb, seems so variable in application, especially in that very trying expression of which we weary so of late — herbaceous border. This seems an architect's phrase rather than a florist's; you always find it on the plans of fine houses with gardens. To me it annihilates every possibility of sentiment, and it usually isn't correct, since many of the plants in these borders are woody perennials instead of annuals; any garden planting that is not "bedding-out" is wildly named "an herbaceous border."

Herb gardens were no vanity and no luxury in our grandmothers' day; they were a necessity. To them every good housewife turned for nearly all that gave variety to her cooking, and to fill her domestic pharmacopœia. The physician placed his chief reliance for supplies on herb gardens and the simples of the fields. An old author says, "Many an old wife or country woman doth often more good with a few known and common garden herbs, than our bombast physicians, with all their prodigious, sumptuous, far-fetched, rare, conjectural medicines." Doctor and goodwife both had a rival in the parson. The picture of the country parson and his wife given by old George Herbert was equally true of the New England minister and his wife: —

> "In the knowledge of simples one thing would be carefully observed, which is to know what herbs may be used instead of drugs of the same nature, and to make the garden the shop; for home-bred medicines are both more easy for the parson's purse, and more familiar for all men's bodies. So when the apothecary useth either for loosing Rhubarb, or for binding Bolearmana, the parson useth damask or white Rose for the one, and Plantain, Shepherd's Purse, and Knot-grass for the other; and that with better success. As for spices, he doth not only prefer home-bred things before them, but condemns them for vanities, and so shuts them out of his family, esteeming that there is no spice comparable for herbs to Rosemary, Thyme, savory Mints, and for seeds to Fennel and Caraway. Accordingly, for

salves, his wife seeks not the city, but prefers her gardens and fields before all outlandish gums."

Simples were medicinal plants, so called because each of these vegetable growths was held to possess an individual virtue, to be an element, a simple substance constituting a single remedy. The noun was generally used in the plural.

You must not think that sowing, gathering, drying, and saving these herbs and simples in any convenient or unstudied way was all that was necessary. Not at all; many and manifold were the rules just when to plant them, when to pick them, how to pick them, how to dry them, and even how to keep them. Gervayse Markham was very wise in herb lore, in the suited seasons of the moon, and hour of the day or night, for herb culling. In the garret of every old house, such as that of the Ward Homestead, shown on page 116, with the wreckage of house furniture, were hung bunches of herbs and simples, waiting for winter use.

The still-room was wholly devoted to storing these herbs and manufacturing their products. This was the careful work of the house mistress and her daughters. It was not intrusted to servants. One book of instruction was entitled, *The Vertuouse Boke of Distyllacyon of the Waters of all Manner of Herbs.* Thomas Tusser wrote: —

"Good huswives provide, ere an sickness do come,
Of sundrie good things in house to have some,
Good aqua composita, vinegar tart,
Rose water and treacle to comfort the heart,

Good herbes in the garden for agues that burn,
That over strong heat to good temper turn."

Both still-room and simple-closet of a dame of the time of Queen Elizabeth or Queen Anne had

Under the Garret Eaves of the Ward Homestead, Shrewsbury, Massachusetts.

crowded shelves. Many an herb and root, unused to-day, was deemed then of sovereign worth. From a manuscript receipt book I have taken names of ingredients, many of which are seldom, perhaps never, used now in medicine. Unripe Blackberries, Ivy berries, Eglantine berries, "Ashen Keys," Acorns, stones of Sloes, Parsley seed, Houseleeks, unripe Hazelnuts, Daisy roots, Strawberry "strings," Woodbine tops, the inner bark of Oak and of red Filberts, green "Broom Cod," White Thorn berries, Turnips, Barberry bark, Dates, Goldenrod, Gourd seed, Blue Lily roots, Parsnip seed, Asparagus roots, Peony roots.

From herbs and simples were made, for internal use, liquid medicines such as wines and waters, syrups, juleps; and solids, such as conserves, confections, treacles, eclėgms, tinctures. There were for external use, amulets, oils, ointments, liniments, plasters, cataplasms, salves, poultices; also sacculi, little bags of flowers, seeds, herbs, etc., and pomanders and posies.

That a certain stimulus could be given to the brain by inhaling the scent of these herbs will not be doubted, I think, by the herb lover even of this century. In the *Haven of Health*, 1636, cures were promised by sleeping on herbs, smelling of them, binding the leaves on the forehead, and inhaling the vapors of their boiling or roasting. Mint was "a good Posie for Students to oft smell." Pennyroyal "quickened the brain by smelling oft." Basil cleared the wits, and so on.

The use of herbs in medicine is far from being obsolete; and when we give them more stately names we swallow the same dose. Dandelion bitters is still used for diseases caused by an ill-working liver.

Wintergreen, which was universally made into tea or oil for rheumatism, appears now in prescriptions for the same disease under the name of Gaultheria. Peppermint, once a sovereign cure for heartburn and "nuralogy," serves us decked with the title of Menthol. "Saffern-tea" never has lost its good standing as a cure for the "jarnders." In country communities scores of old herbs and simples are used in vast amounts; and in every village is some aged man or woman wise in gathering, distilling, and compounding these "potent and parable medicines," to use Cotton Mather's words. One of these gatherers of simples is shown opposite page 120, a quaint old figure, seen afar as we drive through country by-roads, as she bends over some dense clump of weeds in distant meadow or pasture.

In our large city markets bunches of sweet herbs are still sold; and within a year I have seen men passing my city home selling great bunches of Catnip and Mint, in the spring, and dried Sage, Marjoram, and other herbs in the autumn. In one case I noted that it was the same man, unmistakably a real countryman, whom I had noted selling quail on the street, when he had about forty as fine quail as I ever saw. I never saw him sell quail, nor herbs. I think his customers are probably all foreigners — emigrants from continental Europe, chiefly Poles and Italians.

The use of herbs as component parts of love philters and charms is a most ancient custom, and lingered into the nineteenth century in country communities. I knew but one case of the manufacture and administering of a love philter, and it was by a person to whom such an action would seem utterly incongruous. A very gentle, retiring girl in a New England town eighty years ago was deeply in love with the minister whose church she attended, and of which her father was the deacon. The parson was a widower, nearly of middle age, and exceedingly sombre and reserved in character — saddened, doubtless, by the loss of his two young children and his wife through that scourge of New England, consumption; but he was very handsome, and even his sadness had its charm. His house, had burned down as an additional misfortune, and he lived in lodgings with two elderly women of his congregation. Therefore church meetings and various gatherings of committees were held at the deacon's house, and the deacon's daughter saw him day after day, and grew more desperately in love. Desperate certainly she was when she dared even to think of giving a love philter to a minister. The recipe was clearly printed on the last page of an old dream book; and she carried it out in every detail. It was easy to introduce it into the mug of flip which was always brewed for the meeting, and the parson drank it down abstractedly, thinking that it seemed more bitter than usual, but showing no sign of this thought. The philter was promised to have effect in making the drinker love profoundly the first person of opposite sex whom he or she saw after drinking it; and of course the minister saw Hannah as she stood waiting for his empty tankard. The dull details of parish work were talked over in the usual dragging way for half an hour, when the minister became conscious of an intense coldness which seemed to benumb him in every limb; and he tried to walk to the fireplace. Suddenly all in the room became aware that he was very ill, and one called out, "He's got a stroke." Luckily the town doctor was also a deacon, and was therefore present; and he promptly said, "He's poisoned," and hot water from the teakettle, whites of eggs, mustard, and other domestic antidotes were administered with promptitude and effect. It is useless to detail the days of agony to the wretched girl, during which the sick man wavered between life and death, nor her devoted care of him. Soon after his recovery he solemnly proposed marriage to her, and was refused. But he never wavered in his love for her; and every year he renewed his offer and told his wishes, to be met ever with a cold refusal, until ten years had passed; when into his brain there entered a perception that her refusal had some extraordinary element in it. Then, with a warmth of determination worthy a younger man, he demanded an explanation, and received a confession of the poisonous love philter. I suppose time had softened the memory of his suffering, at any rate they were married—so the promise of the love charm came true, after all.

Amos Bronson Alcott was another author of Concord, a sweet philosopher whom I shall ever remember with deepest gratitude as the only person who in my early youth ever imagined any literary capacity in me (and in that he was sadly mistaken, for he fancied I would be a poet). I have read

A Gatherer of Simples.

very faithfully all his printed writings, trying to believe him a great man, a seer; but I cannot, in spite of my gratitude for his flattering though unfulfilled prophecy, discover in his books any profound signs of depth or novelty of thought. In his *Tablets* are some very pleasant, if not surprisingly wise, essays on domestic subjects; one, on "Sweet Herbs," tells cheerfully of the womanly care of the herb garden, but shows that, when written — about 1850 — borders of herbs were growing infrequent.

One great delight of old English gardens is never afforded us in New England; we do not grow Lavender beds. I have of course seen single plants of Lavender, so easily winter-killed, but I never have seen a Lavender bed, nor do I know of one. It is a great loss. A bed or hedge of Lavender is pleasing in the same way that the dress of a Quaker lady is pleasing; it is reposeful, refined. It has a soft effect at the edge of a garden, like a blue-gray haze, and always reminds me of doves. The power of association or some inherent quality of the plant, makes Lavender always suggest freshness and cleanliness.

We may linger a little with a few of these old herb favorites. One of the most balmy and beautiful of all the sweet breaths borne by leaves or blossoms is that of Basil, which, alas! I see so seldom. I have always loved it, and can never pass it without pressing its leaves in my hand; and I cannot express the satisfaction, the triumph, with which I read these light-giving lines of old Thomas Tusser, which showed me why I loved it: —

> "Faire Basil desireth it may be hir lot
> To growe as the gilly flower trim in a pot
> That Ladies and Gentils whom she doth serve
> May help hir as needeth life to preserve."

An explanation of this rhyme is given by *Tusser Redivivus:* "Most people stroak Garden Basil which leaves a grateful smell on the hand and he will have it that Stroaking from a fair lady preserves the life of the Basil."

This is a striking example of floral telepathy; you know what the Basil wishes, and the Basil knows and craves your affection, and repays your caress with her perfume and growth. It is a case of mutual attraction; and I beg the "Gentle Reader" never to pass a pot or plant of Basil without "stroaking" it; that it may grow and multiply and forever retain its relations with fair women, as a type of the purest, the most clinging, and grateful love.

One amusing use of Basil (as given in one of my daughter's old Herbals) was intended to check obesity:—

> "TO MAKE THAT A WOMAN SHALL EAT OF NOTHING THAT IS SET UPON THE TABLE:—Take a little green Basil, and when Men bring the Dishes to the Table put it underneath them that the Woman perceive it not; so Men say that she will eat of none of that which is in the Dish whereunder the Basil lieth."

I cannot understand why so sinister an association was given to a pot of Basil by Boccaccio, who makes the unhappy Isabella conceal the head of her murdered lover in a flower pot under a plant of

Basil; for in Italy Basil is ever a plant of love, not of jealousy or crime. One of its common names is *Bacia, Nicola* — Kiss me, Nicholas. Peasant girls always place Basil in their hair when they go to meet their sweethearts, and an offered sprig of Basil is a love declaration. It is believed that Boccaccio obtained this tale from some tradition of ancient Greece, where Basil is a symbol of hatred and despair. The figure of poverty was there associated with a Basil plant as with rags. It had to be sown with abuse, with cursing and railing, else it would not flourish. In India its sanctity is above all other herbs. A pious Indian has at death a leaf of Basil placed in his bosom as his reward. The house surrounded by Basil is blessed, and all who cherish the plant are sure of heaven.

Mithridate was a favorite medicine of our Puritan ancestors; there were various elaborate compound rules for its manufacture, in which Rue always took a part. It was simple enough in the beginning, when King Mithridates invented it as an antidote against poison: twenty leaves of Rue pounded with two Figs, two dried Walnuts and a grain of salt; which receipt may be taken *cum grano salis.* Rue also entered into the composition of the famous "Vinegar of the Four Thieves." These four rascals, at the time of the Plague in Marseilles, invented this vinegar, and, protected by its power, entered infected houses and carried away property without taking the disease. Rue had innumerable virtues. Pliny says eighty-four remedies were made of it. It was of special use in case of venomous bites, and to counteract "Head-Ach" from over indulgence in wine, especially if a little Sage were added. It promoted love in man and diminished it in woman; it was good for the ear-ache, eye-ache, stomach-ache, leg-ache, back-ache; good for an ague, good for a surfeit; indeed, it would seem wise to make Rue a daily article of food and thus insure perpetual good health.

The scent of Rue seems never dying. A sprig of it was given me by a friend, and it chanced to lie for a single night on the sheets of paper upon which this chapter is written. The scent has never left them, and indeed the odor of Rue hangs literally around this whole book.

Summer Savory and Sweet Marjoram are rarely employed now in American cooking. They are still found in my kitchen, and are used in scant amount as a flavoring for stuffing of fowl. Many who taste and like the result know not the old-fashioned materials used to produce that flavor, and "of the younger sort" the names even are wholly unrecognized.

Sage is almost the only plant of the English kitchen garden which is ordinarily grown in America. I like its fresh grayness in the garden. In the days of our friend John Gerarde, the beloved old herbalist, there was no fixed botanical nomenclature; but he scarcely needed botanical terms, for he had a most felicitous and dextrous use of words. "Sage hath broad leaves, long, wrinkled, rough, and whitish, like in roughness to woollen cloth threadbare." What a description! it is far more vivid than the picture here shown. Sage has never lost its estab-

Our Friend, John Gerarde.

lished place as a flavoring for the stuffing for ducks, geese, and for sausages; but its universal employment as a flavoring for Sage cheese is nearly obsolete. In my childhood home, we always had Sage cheese with other cheeses; it was believed to be an aid in digestion. I had forgotten its taste; and I must say I didn't like it when I ate it last summer, in New Hampshire.

Tansy was highly esteemed in England as a medicine, a cosmetic, and a flavoring and ingredient in cooking. It was rubbed over raw meat to keep the flies away and prevent decay, for in those days of no refrigerators there had to be strong measures taken for the perservation of all perishable food. Its strong scent and taste would be deemed intolerable to us, who can scarce endure even the milder Sage in any large quantity. A good folk name for it is "Bitter Buttons." Gerarde wrote of Tansy,

Sage.

"In the spring time, are made with the leaves hereof newly sprung up, and with Eggs, cakes or Tansies, which be pleasant in Taste and goode for the Stomach."

"To Make a Tansie the Best Way," I learn from *The Accomplisht Cook*, was thus: —

"Take twenty Eggs, and take away five whites, strain them with a quart of good sweet thick Cream, and put to it a grated nutmeg, a race of ginger grated, as much cinnamon beaten fine, and a penny white loaf grated also, mix them all together with a little salt, then stamp some green wheat with some tansie herbs, strain it into the cream and eggs and stir all together; then take a clean frying-pan, and a quarter of a pound of butter, melt it, and put in the tansie, and stir it continually over the fire with a slice, ladle, or saucer, chop it, and break it as it thickens, and being well incorporated put it out of the pan into a dish, and chop it very fine; then make the frying-pan very clean, and put in some more butter, melt it, and fry it whole or in spoonfuls; being finely fried on both sides, dish it up and sprinkle it with rose-vinegar, grape-verjuyce, elder-vinegar, cowslip-vinegar, or the juyce of three or four oranges, and strow on a good store of fine sugar."

To all of this we can say that it would certainly be a very good dish — without the Tansy. Another mediæval recipe was of Tansy, Feverfew, Parsley, and Violets mixed with eggs, fried in butter, and sprinkled with sugar.

The Minnow-Tansie of old Izaak Walton, a "Tanzie for Lent," was made thus: —

"Being well washed with salt and cleaned, and their heads and tails cut off, and not washed after, they prove excellent for that use; that is being fried with the yolks of eggs, the flowers of cowslips and of primroses, and a little tansy, thus used they make a dainty dish."

The name Tansy was given afterward to a rich fruit cake which had no Tansy in it. It was apparently a favorite dish of Pepys. A certain derivative custom obtained in some New England towns — certainly in Hartford and vicinity. Tansy was used to flavor the Fast Day pudding. One old lady recalls that it was truly a bitter food to the younger members of the family; Miss Shelton, in her entertaining book, *The Salt Box House*, tells of Tansy cakes, and says children did not dislike them. Tansy bitters were made of Tansy leaves placed in a bottle with New England rum. They were a favorite spring tonic, where all physicians and housewives prescribed "the bitter principle" in the spring time.

No doubt Tansy was among the earliest plants brought over by the settlers; it was carefully cherished in the herb garden, then spread to the dooryard and then to farm lanes. As early as 1746 the traveller Kalm noted Tansy growing wild in hedges and along roads in Pennsylvania. Now it extends its sturdy growth for miles along the country road, one of the rankest of weeds. It still is used in the manufacture of proprietary medicines, and for this purpose is cut with a sickle in great armfuls and gathered in cartloads. I have always liked its scent; and its leaves, as Gerarde said, "infinitely jagged and'nicked and curled"; and its cheerful little "bitter buttons" of gold. Some old flowers adapt themselves to modern conditions and look up-to-date; but to me the Tansy, wherever found, is as openly old-fashioned as a betty-lamp or a foot-stove.

On July 1, 1846, an old grave was opened in the ancient "God's Acre" near the halls of Harvard University in Cambridge, Massachusetts. This grave was a brick vault covered with irregularly shaped flagstones about three inches thick. Over it was an ancient slab of peculiar stone, unlike any others in the cemetery save those over the graves

Tansy.

of two presidents of the College, Rev. Dr. Chauncy and Dr. Oakes. As there were headstones near this slab inscribed with the names of the great-grandchildren of President Dunster, it was believed that this was the grave of a third President, Dr. Dunster. He died in the year 1659; but his death took place in midwinter; and when this coffin was opened, the skeleton was found entirely surrounded with common Tansy, in seed, a portion of which had been pulled up by the roots, and it was therefore believed by many who thought upon the matter that it was the coffin and grave of President Mitchell, who died in July, 1668, of "an extream fever." The skeleton was found still wrapped in a cerecloth, and in the record of the church is a memorandum of payment "for a terpauling to wrap Mr. Mitchell." The Tansy found in this coffin, placed there more than two centuries ago, still retained its shape and scent.

This use of Tansy at funerals lingered long in country neighborhoods in New England, in some vicinities till fifty years ago. To many older persons the Tansy is therefore so associated with grewsome sights and sad scenes, that they turn from it wherever seen, and its scent to them is unbearable. One elderly friend writes me: "I never see the leaves of Tansy without recalling also the pale dead faces I have so often seen encircled by the dank, ugly leaves. Often as a child have I been sent to gather all the Tansy I could find, to be carried by my mother to the house of mourning; and I gathered it, loathing to touch it, but not daring to refuse, and I loathe it still."

Tansy not only retains its scent for a long period, but the "golden buttons" retain their color; I have

Garden of Mrs. Abraham Lansing, Albany, New York.

seen them in New England parlors forming part of a winter posy; this, I suppose, in neighborhoods where Tansy was little used at funerals.

If an herb garden had no other reason for existence, let me commend it to the attention of those of ample grounds and kindly hearts, for a special purpose — as a garden for the blind. Our many flower-charities furnish flowers throughout the summer to our hospitals, but what sweet-scented flowers are there for those debarred from any sight of beauty? Through the past summer my daughters sent several times a week, by the generous carriage of the Long Island Express Company, boxes of wild flowers to any hospital of their choice. What could we send to the blind? The midsummer flowers of field and meadow gratified the sight, but scent was lacking. A sprig of Sweet Fern or Bayberry was the only resource. Think of the pleasure which could be given to the sightless by a posy of sweet-scented leaves, by Southernwood, Mint, Balm, or Basil, and when memory was thereby awakened in those who once had seen, what tender thoughts! If this book could influence the planting of an herb garden for the solace of those who cannot see the flowers of field and garden, then it will not have been written in vain.

CHAPTER VI

IN LILAC TIDE

"Ere Man is aware
That the Spring is here
The Flowers have found it out."

—*Ancient Chinese Saying.*

A FLOWER opens, and lo! another Year," is the beautiful and suggestive legend on an old vessel found in the Catacombs. Since these words were written, how many years have begun! how many flowers have opened! and yet nature has never let us weary of spring and spring flowers. My garden knows well the time o' the year. It needs no almanac to count the months.

"The untaught Spring is wise
In Cowslips and Anemonies."

While I sit shivering, idling, wondering when I can "start the garden"—lo, there are Snowdrops and spring starting up to greet me.

Ever in earliest spring are there days when there is no green in grass, tree, or shrub; but when the garden lover is conscious that winter is gone and spring is waiting. There is in every garden, in every dooryard, as in the field and by the roadside, in some indefinable way a look of spring. One hint of spring comes even before its flowers — you can smell its coming. The snow is gone from the garden walks and some of the open beds; you walk warily down the softened path at midday, and you smell the earth as it basks in the sun, and a

Ladies' Delights.

faint scent comes from some twigs and leaves. Box speaks of summer, not of spring; and the fragrance from that Cedar tree is equally suggestive of summer. But break off that slender branch of Calycanthus — how fresh and welcome its delightful spring scent. Carry it into the house with branches of Forsythia, and how quickly one fills its leaf buds and the other blossoms.

For several years the first blossom of the new year in our garden was neither the Snowdrop nor Crocus, but the Ladies' Delight, that laughing, speaking little garden face, which is not really a spring flower, it is a stray from summer; but it is such a shrewd, intelligent little creature that it readily found out that spring was here ere man or other flowers knew it. This dear little primitive of the Pansy tribe has become wonderfully scarce save in cherished old gardens like those of Salem, where I saw this year a space thirty feet long and several feet wide, under flowering shrubs and bushes, wholly covered with the everyday, homely little blooms of Ladies' Delights. They have the party-colored petal of the existing strain of English Pansies, distinct from the French and German Pansies, and I doubt not are the descendants of the cherished garden children of the English settlers. Gerarde describes this little English Pansy or Heartsease in 1587 under the name of *Viola tricolor*: —

"The flouers in form and figure like the Violet, and for the most part of the same Bignesse, of three sundry colours, purple, yellow and white or blew, by reason of the beauty and braverie of which colours they are very pleasing to the eye, for smel they have little or none."

In Breck's *Book of Flowers*, 1851, is the first printed reference I find to the flower under the name Ladies' Delight. In my childhood I never heard it called aught else; but it has a score of folk names, all testifying to an affectionate intimacy: Bird's-eye; Garden-gate; Johnny-jump-up; None-

Garden House in Garden of Hon. William H. Seward, Auburn, New York.

so-pretty; Kitty-come; Kit-run-about; Three-faces under-a-hood; Come-and-cuddle-me; Pink-of-my Joan; Kiss-me; Tickle-my-fancy; Kiss-me-ere-I rise; Jump-up-and-kiss-me. To our little flower has also been given this folk name, Meet-her-in-the-entry-kiss-her-in-the-buttery, the longest plant name in the English language, rivalled only by Miss Jekyll's triumph of nomenclature for the Stone-crop, namely: Welcome-home-husband-be-he-ever-so-drunk.

These little Ladies' Delights have infinite variety of expression; some are laughing and roguish, some sharp and shrewd, some surprised, others worried, all are animated and vivacious, and a few saucy to a degree. They are as companionable as people — nay, more; they are as companionable as children. No wonder children love them; they recognize kindred spirits. I know a child who picked unbidden a choice Rose, and hid it under her apron. But as she passed a bed of Ladies' Delights blowing in the wind, peering, winking, mocking, she suddenly threw the Rose at them, crying out pettishly, "Here! take your old flower!"

The Dandelion is to many the golden seal of spring, but it blooms the whole circle of the year in sly garden corners and in the grass. Of it might have been written the lines: —

> "It smiles upon the lap of May,
> To sultry August spreads its charms,
> Lights pale October on its way,
> And twines December's arms."

I have picked both Ladies' Delights and Dandelions every month in the year.

I suppose the common Crocus would not be deemed a very great garden ornament in midsummer, in its lowly growth; but in its spring blossom-

Sun-dial in Garden of Hon. William H. Seward, Auburn, New York.

ing it is — to use another's words — "most gladsome of the early flowers." A bed of Crocuses is certainly a keen pleasure, glowing in the sun, almost as grateful to the human eye as to the honey-gathering bees that come unerringly, from somewhere, to hover over the golden cups. How welcome after winter is the sound of that humming.

In the garden's story, there are ever a few pictures which stand out with startling distinctness. When the year is gone you do not recall many days nor many flowers with precision; often a single flower seems of more importance than a whole garden. In the day book of 1900 I have but few pictures; the most vivid was the very first of the season. It could have been no later than April, for one or two Snowdrops still showed white in the grass, when a splendid ribbon of Chionodoxa — Glory of the Snow — opened like blue fire burning from plant to plant, the bluest thing I ever saw in any garden. It was backed with solid masses of equally vivid yellow Alyssum and chalk-white Candy-tuft, both of which had had a good start under glass in a temporary forcing bed. These three solid masses of color surrounded by bare earth and showing little green leafage made my eyes ache, but a picture was burnt in which will never leave my brain. I always have a sense of importance, of actual ownership of a plant, when I can recall its introduction — as I do of the Chionodoxa, about 1871. It is said to come up and bloom in the snow, but I have never seen it in blossom earlier than March, and never then unless the snow has vanished. It has much of the charm of its relative, the Scilla.

We all have flower favorites, and some of us have flower antipathies, or at least we are indifferent to certain flowers; but I never knew any one but loved the Daffodil. Not only have poets and dramatists sung it, but it is a common favorite, as shown by its homely names in our everyday speech. I am always touched in *Endymion* that the only flowers named as "a thing of beauty that is a joy forever" are Daffodils "with the green world they live in."

In Daffodils I like the "old fat-headed sort with nutmeg and cinnamon smell and old common English names — Butter-and-eggs, Codlins-and-cream, Bacon and eggs." The newer ones are more slender in bud and bloom, more trumpet-shaped, and are commonplace of name instead of common. In Virginia the name of a variety has become applied to a family, and all Daffodils are called Butter-and-eggs by the people.

On spring mornings the Tulips fairly burn with a warmth, which makes them doubly welcome after winter. Emerson — ever able to draw a picture in two lines — to show the heart of everything in a single sentence — thus paints them : —

> "The gardens fire with a joyful blaze
> Of Tulips in the morning's rays."

"Tulipase do carry so stately and delightful a form, and do abide so long in their bravery, that there is no Lady or Gentleman of any worth that is not caught with this delight," — wrote the old herbalist Parkinson. Bravery is an ideal expression for Tulips.

It is with something of a shock that we read the words of Philip Hamerton in *The Sylvan Year*, that nature is not harmonious in the spring, but is only in the way of becoming so. He calls it the time of crudities, like the adolescence of the mind. He says,

Lilacs in Midsummer in Garden of Mrs. Abraham Lansing,
Albany, New York.

"The green is good for us, and we welcome it with uncritical gladness; but when we think of painting, it may be doubted whether any season of the year is less propitious to the broad and noble harmonies which are the secrets of all grand effects in art." And he compares the season to the uncomfortable hour in a household when the early risers are walking about, not knowing what to do with themselves, while others have not yet come down to breakfast.

I must confess that an undiversified country landscape in spring has upon me the effect asserted by Hamerton. I recall one early spring week in the Catskills, when I fairly complained, "Everything is so green here." I longed for rocks, water, burnt fields, bare trees, anything to break that glimmering green of new grass and new Birches. But in the spring garden there is variety of shape and color; the Peony leaf buds are red, some sprouting leaves are pink, and there are vast varieties of brown and gray and gold in leaf.

Let me give the procession of spring in the garden in the words of a lover of old New England flowers, Dr. Holmes. It is a vivid word picture of the distinctive forms and colors of budding flowers and leaves.

"At first the snowdrop's bells are seen,
  Then close against the sheltering wall
The tulip's horn of dusky green,
  The peony's dark unfolding ball.

"The golden-chaliced crocus burns;
  The long narcissus blades appear;

The cone-beaked hyacinth returns
To light her blue-flamed chandelier.

"The willow's whistling lashes, wrung
By the wild winds of gusty March,
With sallow leaflets lightly strung,
Are swaying by the tufted larch.

"See the proud tulip's flaunting cup,
That flames in glory for an hour, —
Behold it withering, then look up —
How meek the forest-monarchs flower!

"When wake the violets, Winter dies;
When sprout the elm buds, Spring is near;
When lilacs blossom, Summer cries,
'Bud, little roses, Spring is here.'"

The universal flower in the old-time garden was the Lilac; it was the most beloved bloom of spring, and gave a name to Spring — Lilac tide. The Lilac does not promise "spring is coming"; it is the emblem of the *presence* of spring. Dr. Holmes says, "When Lilacs blossom, Summer cries, '*Spring is here*'" in every cheerful and lavish bloom. Lilacs shade the front yard; Lilacs grow by the kitchen doorstep; Lilacs spring up beside the barn; Lilacs shade the well; Lilacs hang over the spring house; Lilacs crowd by the fence side and down the country road. In many colonial dooryards it was the only shrub — known both to lettered and unlettered folk as Laylock, and spelt Laylock too. Walter Savage Landor, when Laylock had become antiquated, still clung to the word, and used it with a stubborn persistence such as he alone could compass, and which seems strange in the most finished classical scholar of his day.

"I shall not go to town while the Lilacs bloom," wrote Longfellow; and what Lilac lover could have

Lilacs at Craigie House, the Home of Longfellow.

left a home so Lilac-embowered as Craigie House! A view of its charms in Lilac tide is given in outline on this page; the great Lilac trees seem wondrously suited to the fine old Revolutionary mansion.

There is in Albany, New York, a lovely garden endeared to those who know it through the memory of a presence that lighted all places associated with it with the beauty of a noble life. It is the garden of the home of Mrs. Abraham Lansing, and was planted by her father and mother, General

Box-edged Garden at the Home of Longfellow.

and Mrs. Peter Gansevoort, in 1846, having been laid out with taste and an art that has borne the test of over half a century's growth. In the garden are scores of old-time favorites: Flower de Luce, Peonies, Daffodils, and snowy Phlox; but instead of bending over the flower borders, let us linger awhile in the wonderful old Lilac walk. It is a glory of tender green and shaded amethyst and grateful hum of bees, the very voice of Spring. Every sense is gratified, even that of touch, when the delicate plumes of the fragrant Lilac blossoms brush your cheek as you walk through its path; there is no spot of fairer loveliness than this Lilac walk in May. It is a wonderful study of flickering light and grateful shade in midsummer. Look at its full-leaf charms opposite page 138; was there ever anything lovelier in any garden, at any time, than the green vista of this Lilac walk in July? But for the thoughtful garden-lover it has another beauty still, the delicacy and refinement of outline when the Lilac walk is bare of foliage, as is shown on page 220 and facing page 154. The very spirit of the Lilacs seems visible, etched with a purity of touch that makes them sentient, speaking beings, instead of silent plants. See the outlines of stem and branch against the tender sky of this April noon. Do you care for color when you have such beauty of outline? Surely this Lilac walk is loveliest in April, with a sensitive etherealization beyond compare. How wonderfully these pictures have caught the look of tentative spring—spring waiting for a single day to burst into living green. There is an ancient Saxon name for springtime — Opyn-tide — thus defined by an old writer, "Whenne that flowres think on blowen"—when the flowers begin to think of budding and blowing; and so I name this picture Opyn-tide, the Thought of Spring.

For many years Lilacs were planted for hedges;

they were seldom satisfactory if clipped, for the broad-spreading leaves were always gray with dust, and they often had a "rust" which wholly destroyed their beauty. The finest clipped Lilac hedge I ever saw is at Indian Hill, Newburyport. It was set out about 1850, and is compact and green as Privet; the leaves are healthy, and the growth perfect down to the ground; it is an unusual example of Lilac growth — a perfect hedge. An unclipped Lilac hedge is lovely in its blooming; a beautiful one grows by the side of the old family home of Mr. Mortimer Howell at West Hampton Beach, Long Island. To this hedge in May come a-begging dusky city flower venders, who break off and carry away wagon loads of blooms. As the fare from and to New York is four dollars, and a wagon has to be hired to convey the flowers from the hedge two miles to the railroad station, there must be a high price charged for these Lilacs to afford any profit; but the Italian flower sellers appear year after year.

Lilacs bloom not in our ancient literature; they are not named by Shakespeare, nor do I recall any earlier mention of them than in the essay of Lord Bacon on "Gardens," published about 1610, where he spelled it Lelacke. Blue-pipe tree was the ancient name of the Lilac, a reminder of the time when pipes were made of its wood; I heard it used in modern speech once. An old Narragansett coach driver called out to me, "Ye set such store on flowers, don't ye want to pick that Blue-pipe in Pender Zeke's garden?" — a deserted garden and home at Pender Zeke's Corner. This man had some of the traits of Mrs. Wright's delightful "Time-o'-Day," and he knew well my love of flowers; for he had been my charioteer to the woods where Rhododendron and Rhodora bloom, and he had revealed to me the pond where grew the pink Water Lilies. And from a chance remark of mine he had conveyed to me a wagon load of Joepye-weed and Boneset, to the dismay of my younger children, who had apprehensions of unlimited gallons of herb tea therefrom. Let me steal a few lines from my spring Lilacs to write of these two "Sisters of Healing," which were often planted in the household herb garden. From July to September in the low lying meadows of every state from the Bay of Fundy to the Gulf of Mexico, can be found Joepye-weed and Boneset. The dull pink clusters of soft fringy blooms of Joepye-weed stand up three to eight feet in height above the moist earth, catching our eye and the visit of every passing butterfly, and commanding attention for their fragrance, and a certain dignity of carriage notable even among the more striking hues of the brilliant

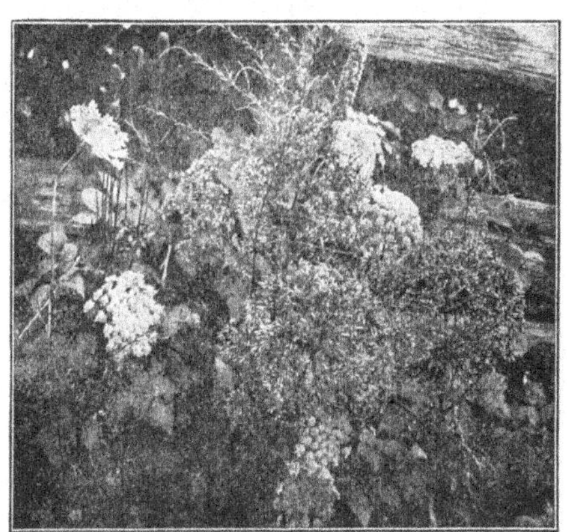
Joepye-weed and Queen Anne's Lace.

Goldenrod and vivid Sunflowers. Joe Pye was an Indian medicine-man of old New England, famed among his white neighbors for his skill in curing the devastating typhoid fevers which, in those days of no drainage and ignorance of sanitation, vied with so-called "hereditary" consumption in exterminating New England families. His cure-all was a bitter tea decocted from leaves and stalks of this *Eupatorium purpureum*, and in token of his success the plant bears everywhere his name, but it is now wholly neglected by the simpler and herb-doctor. The sister plant, the *Eupatorium perfoliatum*, known as Thoroughwort, Boneset, Ague-weed, or Indian Sage, grows everywhere by its side, and is also used in fevers. It was as efficacious in "break bone fever" in the South a century ago as it is now for the grippe, for it still is used, North and South, in many a country

Boneset.

home. Neltje Blanchan and Mrs. Dana Parsons call Thoroughwort or Boneset tea a "nauseous draught," and I thereby suspect that neither has tasted it. I have many a time, and it has a clear, clean bitter taste, no stronger than any bitter beer or ale. Every year is Boneset gathered in old Narragansett; but swamp edges and meadows that are easy of access have been depleted of the stately growth of saw-edged wrinkled leaves, and the Boneset gatherer must turn to remote brooksides and inaccessible meadows for his harvest. The flat-topped terminal cymes of leaden white blooms are not distinctive as seen from afar, and many flowers of similar appearance lure the weary simpler here and there, until at last the welcome sight of the connate perfoliate leaves, surrounding the strong stalk, distinctive of the Boneset, show that his search is rewarded.

After these bitter draughts of herb tea, we will turn, as do children, to sweets, to our beloved Lilac blooms. The Lilac has ever been a flower welcomed by English-speaking folk since it first came to England by the hand of some mariner. It is said that a German traveller named Busbeck brought it from the Orient to the continent in the sixteenth century. I know not when it journeyed to the new world, but long enough ago so that it now grows cheerfully and plentifully in all our states of temperate clime and indeed far south. It even grows wild in some localities, though it never looks wild, but plainly shows its escape or exile from some garden. It is specially beloved in New England, and it seems so much more suited in spirit to New England than to

Persia that it ought really to be a native plant. Its very color seems typical of New England; some parts of celestial blue, with more of warm pink, blended and softened by that shading of sombre gray ever present in New England life into a distinctive color known everywhere as lilac — a color grateful, quiet, pleasing, what Thoreau called a "tender, civil, cheerful color." Its blossoming at the time of Election Day, that all-important New England holiday, gave it another New England significance.

There is no more emblematic flower to me than the Lilac; it has an association of old homes, of home-making and home interests. On the country farm, in the village garden, and in the city yard, the lilac was planted wherever the home was made, and it attached itself with deepest roots, lingering sometimes most sadly but sturdily, to show where the home once stood.

Let me tell of two Lilacs of sentiment. One of them is shown on page 149; a glorious Lilac tree which is one of a group of many full-flowered, pale-tinted ones still growing and blossoming each spring on a deserted homestead in old Narragansett. They bloom over the grave of a fine old house, and the great chimney stands sadly in their midst as a gravestone. "Hopewell," ill-suited of name, was the home of a Narragansett Robinson famed for good cheer, for refinement and luxury, and for a lovely garden, laid out with cost and care and filled with rare shrubs and flowers. Perhaps these Lilacs were a rare variety in their day, being pale of tint;

Magnolias.

now they are as wild as their companions, the Cedar hedges.

Gathering in the front dooryard of a fallen farm-house some splendid branches of flowering Lilac, I found a few feet of cellar wall and wooden house side standing, and the sills of two windows. These window sills, exposed for years to the bleaching and

Lilacs at Hopewell.

fading of rain and sun and frost, still bore the circular marks of the flower pots which, filled with house-plants, had graced the kitchen windows for many a winter under the care of a flower-loving house mistress. A few days later I learned from a woman over ninety years of age — an inmate of the "Poor House" — the story of the home thus touchingly indicated by the Lilac bushes and the stains of the flower pots. Over eighty years ago she had brought the tiny Lilac-slip to her childhood's home, then standing in a clearing in the forest. She carried it carefully in her hands as she rode behind her father on a pillion after a visit to her grandmother. She and her little brothers and sisters planted the tiny thing "of two eyes only," as she said, in the shadow of the house, in the little front yard. And these children watered it and watched it, as it rooted and grew, till the house was surrounded each spring with its vivacious blooms, its sweet fragrance. The puny slip has outlived the house and all its inmates save herself, outlived the brothers and sisters, their children and grandchildren, outlived orchard and garden and field. And it will live to tell a story to every thoughtful passer-by till a second growth of forest has arisen in pasture and garden and even in the cellar-hole, when even then the cheerful Lilac will not be wholly obliterated.

A bunch of early Lilacs was ever a favorite gift to "teacher," to be placed in a broken-nosed pitcher on her desk. And Lilac petals made such lovely necklaces, thrust within each other or strung with needle and thread. And there was a love divination by Lilacs which we children solemnly observed. There will occasionally appear a tiny Lilac flower, usually a white Lilac, with five divisions of the petal instead of four — this is a Luck Lilac. This must be solemnly swallowed. If it goes down smoothly, the dabbler in magic cries out, "He loves me;" if she chokes at her floral food, she must say sadly, "He loves me not." I remember once calling out, with gratification and pride, "He loves me!"

"Who is he?" said my older companions. "Oh, I didn't know he had to be somebody," I answered in surprise, to be met by derisive laughter at my satisfaction with a lover in general and not in particular. It was a matter of Lilac-luck-etiquette that the

Persian Lilacs and Peonies in Garden of the Kimball Homestead, Portsmouth, New Hampshire.

lover's name should be pronounced mentally before the petal was swallowed.

In the West Indies the Lilac is a flower of mysterious power; its perfume keeps away evil spirits, ghosts, banshees. If it grows not in the dooryard, its protecting branches are hung over the doorway. I think of this when I see it shading the door of happy homes in New England.

In our old front yards we had only the common

Lilacs, and occasionally a white one; and as a rarity the graceful, but sometimes rather spindling, Persian Lilacs, known since 1650 in gardens, and shown on page 151. How the old gardens would have stared at the new double Lilacs, which have luxuriant plumes of bloom twenty inches long.

The "pensile Lilac" has been sung by many poets; but the spirit of the flower has been best portrayed in verse by Elizabeth Akers. I can quote but a single stanza from so many beautiful ones.

> "How fair it stood, with purple tassels hung,
> Their hue more tender than the tint of Tyre;
> How musical amid their fragrance rung
> The bee's bassoon, keynote of spring's glad choir!
> O languorous Lilac! still in time's despite
> I see thy plumy branches all alight
> With new-born butterflies which loved to stay
> And bask and banquet in the temperate ray
> Of springtime, ere the torrid heats should be:
> For these dear memories, though the world grow gray,
> I sing thy sweetness, lovely Lilac tree!"

Another poet of the Lilac is Walt Whitman. He tells his delight in "the Lilac tall and its blossoms of mastering odor." He sings: "with the birds a warble of joy for Lilac-time." That noble, heroic dirge, the *Burial Hymn of Lincoln*, begins: —

> "When Lilacs last in the dooryard bloom'd."

The poet stood under the blossoming Lilacs when he learned of the death of Lincoln, and the scent and sight of the flowers ever bore the sad association. In this poem is a vivid description of —

"The Lilac bush, tall growing, with heart-shaped leaves of rich green,
With many a pointed blossom, rising delicate with the perfume strong I love.
With every leaf a miracle."

Thomas William Parsons could turn from his profound researches and loving translations of Dante to write with deep sympathy of the Lilac. His verses have to me an additional interest, since I believe they were written in the house built by my ancestor in 1740, and occupied still by his descendants. In its front dooryard are Lilacs still standing under the windows of Dr. Parsons' room, in which he loved so to write.

Hawthorne felt a sort of "ludicrous unfitness in the idea of a time-stricken and grandfatherly Lilac bush." He was dissatisfied with aged Lilacs, though he knew not whether his heart, judgment, or rural sense put him in that condition. He felt the flower should either flourish in immortal youth or die. Apple trees could grow old and feeble without his reproach, but an aged Lilac was improper.

I fancy no one ever took any care of Lilacs in an old garden. As soon water or enrich the Sumach and Elder growing by the roadside! But care for your Lilacs nowadays, and see how they respond. Make them a *garden* flower, and you will never regret it. There be those who prefer grafted Lilacs — the stock being usually a Syringa; they prefer the single trunk, and thus get rid of the Lilac suckers. But compare a row of grafted Lilacs to a row of natural fastigate growth, as shown on page 220, and I think nature must be preferred.

"Methinks I see my contemplative girl now in the garden watching the gradual approach of Spring," wrote Sterne. My contemplative girl lives in the city, how can she know that spring is here? Even on those few square feet of mother earth, dedicated to clotheslines and posts, spring sets her mark. Our Lilacs seldom bloom, but they put forth lovely fresh green leaves; and even the unrolling of the leaves of our Japanese ivies are a pleasure.

Our poor little strips of back yard in city homes are apt to be too densely shaded for flower blooms, but some things will grow, even there. Some wild flowers will live, and what a delight they are in spring. We have a Jack-in-the-pulpit who comes up just as jauntily there as in the wild woods; Dog-tooth Violet and our common wild Violet also bloom. A city neighbor has Trillium which blossoms each year; our Trillium shows leaves, but no blossoms, and does not increase in spread of roots. Bloodroot, a flower so shy when gathered in the woods, and ever loving damp sites, flourishes in the dryest flower bed, grows coarser in leaf and bloom, and blossoms earlier, and holds faster its snowy petals. Corydalis in the garden seems so garden-bred that you almost forget the flower was ever wild.

The approach of spring in our city parks is marked by the appearance of the Dandelion gatherers. It is always interesting to see, in May, on the closely guarded lawns and field expanses of our city parks, the hundreds of bareheaded, gayly-dressed Italian and Portuguese women and children eagerly gather-

Opyn-tide, the Thought of Spring.

ing the young Dandelion plants to add to their meagre fare as a greatly-loved delicacy. They collect these "greens" in highly-colored kerchiefs, in baskets, in squares of sheeting; I have seen the women bearing off a half-bushel of plants; even their stumpy little children are impressed to increase the welcome harvest, and with a broken knife dig eagerly in the greensward. The thrifty park commissioners, in Dandelion-time, relax their rigid rules, "Keep Off the Grass," and turn the salad-loving Italians loose to improve the public lawns by freeing them from weeds.

The earliest sign of spring in the fields and woods in my childhood was the appearance of the Willow catkins, and was heralded by the cry of one child to another, — "Pussy-willows are out." How eagerly did those who loved the woods and fields turn, after the storm, whiteness, and chill of a New England winter, to Pussy-willows as a promise of summer and sunshine. Some of their charm ever lingers to us as we see them in the baskets of swarthy street venders in New York.

Magnolia blossoms are sold in our city streets to remind city dwellers of spring. "Every flower its own bow-kwet," is the call of the vender. Bunches of Locust blossoms follow, awkwardly tied together. Though the Magnolia is earlier, I do not find it much more splendid as a flowering tree for the garden than our northern Dogwood; and the Dogwood when in bloom seems just as tropical. It is then the glory of the landscape; and its radiant starry blossoms turn into ideal beauty even our sombre cemeteries.

The Magnolia has been planted in northern gardens for over a century. Gardens on Long Island have many beautiful old specimens, doubtless furnished by the Prince Nurseries. These seem thoroughly at home; just as does the Locust brought from Virginia, a century ago, by one Captain Sands of Sands Point, to please his Virginia bride with the presence of the trees of her girlhood's home. These Locusts have spread over every rood of Long Island earth, and seem as much at home as Birch or Willow. The three Magnolia trees on Mr. Brown's lawn in Flatbush are as large as any I know in the North, and were exceptionally full of bloom this year, this photograph (shown facing page 148) being taken when they were past their prime. I saw children eagerly gathering the waxy petals which had fallen, and which show so plainly in the picture. But the flower is not common enough here for northern children to learn the varied attractions of the Magnolia.

The flower lore of American children is nearly all of English derivation; but children invent as well as copy. In the South the lavish growth of the Magnolia affords multiform playthings. The beautiful broad white petals give a snowy surface for the inditing of messages or valentines, which are written with a pin, when the letters turn dark brown. The stamens of the flower — waxlike with red tips — make mock illuminating matches. The leaves shape into wonderful drinking cups, and the scarlet seeds give a glowing necklace.

The glories of a spring garden are not in the

rows of flowering bulbs, beautiful as they are; but in the flowering shrubs and trees. The old garden had few shrubs, but it had unsurpassed beauty in its rows of fruit trees which in their blossoming give the spring garden, as here shown, that lovely

A Thought of Winter's Snows.

whiteness which seems a blending of the seasons — a thought of winter's snows. The perfection of Apple blossoms I have told in another chapter. Earlier to appear was the pure white, rather chilly, blooms of the Plum tree, to the Japanese "the eldest brother of an hundred flowers." They are faintly sweet-scented with the delicacy found in many spring blossoms. A good example of the short verses of the Japanese poets tells of the Plum blossom and its perfume.

> "In springtime, on a cloudless night,
> When moonbeams throw their silver pall
> O'er wooded landscapes, veiling all
> In one soft cloud of misty white,
> 'Twere vain almost to hope to trace
> The Plum trees in their lovely bloom
> Of argent; 'tis their sweet perfume
> Alone which leads me to their place."

The lovely family of double white Plum blossoms which now graces our gardens is varied by tinted ones; there are sixty in all which the nineteenth century owes to Japan.

The Peach tree has a flower which has given name to one of the loveliest colors in the world. The Peach has varieties with wonderful double flowers of glorious color. Cherry trees bear a more cheerful white flower than Plum trees.

> "The Cherry boughs above us spread
> The whitest shade was ever seen;
> And flicker, flicker came and fled
> Sun-spots between."

I do not recall the Judas tree in my childhood. I am told there were many in Worcester; but there were none in our garden, nor in our neighborhood, and that was my world. Orchids might have hung from the trees a mile from my home, and would have been no nearer me than the tropics. I had a small world, but it was large enough, since it was bounded by garden walls.

Almond trees are seldom seen in northern gardens; but the Flowering Almond flourishes as one of the purest and loveliest familiar shrubs. Silvery pink in bloom when it opens, the pink darkens till when in full flower it is deeply rosy. It was, next to the Lilac, the favorite shrub of my childhood. I used to call the exquisite little blooms "fairy roses," and there were many fairy tales relating to the Almond bush. This made the flower enhaloed with sentiment and mystery, which charmed as much as its beauty. The Flowering Almond seemed to have a special place under a window in country yards and gardens, as it is shown on page 39. A fitting spot it was, since it never grew tall enough to shade the little window panes.

With Pussy-willows and Almond blossoms and Ladies' Delights, with blossoming playhouse Apple trees and sweet-scented Lilac walks, spring was certainly Paradise in our childhood. Would it were an equally happy season in mature years; but who, garden-bred, can walk in the springtime through the garden of her childhood without thought of those who cared for the garden in its youth, and shared the care of their children with the care of their flowers, but now are seen no more.

> "Oh, far away in some serener air,
> The eyes that loved them see a heavenly dawn:
> How can they bloom without her tender care?
> Why should they live when her sweet life is gone?"

I have written of the gladness of spring, but I know nothing more overwhelming than the heartache of spring, the sadness of a fresh-growing spring garden. Where is the dear one who planted it and loved it, and he who helped her in the care, and the loving child who played in it and left it in the springtime? All that is good and beautiful has come again to us with the sunlight and warmth, save those whom we still love but can see no more. By that very measure of happiness poured for us in childhood in Lilac tide, is our cup of sadness now filled.

CHAPTER VII

OLD FLOWER FAVORITES

"God does not send us strange flowers every year.
When the spring winds blow o'er the pleasant places
The same dear things lift up the same fair faces;
The Violet is here.

"It all comes back; the odor, grace, and hue
Each sweet relation of its life repeated;
No blank is left, no looking-for is cheated;
It is the thing we knew."

— ADELINE D. T. WHITNEY, 1861.

NOT only do I love to see the same dear things year after year, and to welcome the same odor, grace, and hue; but I love to find them in the same places. I like a garden in which plants have been growing in one spot for a long time, where they have a fixed home and surroundings. In our garden the same flowers shoulder each other comfortably and crowd each other a little, year after year. They look, my sister says, like long-established neighbors, like old family friends, not as if they had just "moved in," and didn't know each other's names and faces. Plants grow better when they are among flower friends. I suppose we have to transplant some plants, sometimes; but I would try to keep old friends together even in those removals. They would be lonely when they opened their eyes after the winter's sleep, and saw strange flower forms and unknown faces around them.

Larkspur and Phlox.

For flowers have friendships, and antipathies as well. How Canterbury Bells and Foxgloves love to grow side by side! And Sweet Williams, with Foxgloves, as here shown. And in my sister's garden Larkspur always starts up by white Phlox — see a bit of the border on this page. Whatever may influence these docile alliances, it isn't a proper sense of fitness of color; for Tiger Lilies dearly love to grow by crimson-purple Phlox, a most inharmonious association, and you can hardly separate them. If a flower dislikes her neighbor in the garden, she moves quietly away, I don't know where or how. Sometimes she dies, but at any rate she is gone. It is so queer; I have tried every year to make Feverfew grow in this bed, and it won't do it, though it grows across the path. There is some flower here that the pompous Feverfew doesn't care to associate with. Not the Larkspur, for they are famous friends — perhaps it is the Sweet William, who is rather a plain fellow. In general flowers are very sociable with each other, but they have some preferences, and these are powerful ones.

Sweet William and Foxglove.

It is amusing to read in no less than five recent English "garden-books," by flower-loving souls, the solemn advice that if you wish a beautiful garden effect you "must plant the great Oriental Poppy by the side of the White Lupine."

"Thou say'st an undisputed thing
In such a solemn way."

The truth is, you have very little to do with it. That Poppy chooses to keep company with the White Lupine, and to that impulse you owe your fine garden effect. The Poppy is the slyest magician of the whole garden. He comes and goes at will. This year a few blooms, nearly all in one corner; next year a blaze of color banded across the middle of the garden like the broad sash of a court chamberlain. Then a single grand blossom quite alone in the pansy bed, while another pushes up between the tight close leaves of the box edging: — the Poppy is *queer*.

Plume Poppy.

Some flowers have such a hatred of man they cannot breathe and live in his presence, others have an equal love of human companionship. The white Clover clings here to our pathway as does the English Daisy across seas. And in our garden Ladies' Delights and Ambrosia tell us, without words, of their love for us and longing to be by our side; just as plainly as a child silently tells us his love and dependence on us by taking our hand as we walk side by side. There is not another gesture of childhood, not an affectionate word which ever touched my heart as did that trustful holding of the hand. One of my children throughout his brief life never walked by my side without clinging closely — I think without conscious intent — with his little hand to mine. I can never forget the affection, the trust of that vanished hand.

I find that my dearest flower loves are the old flowers, — not only old to me because I knew them in childhood, but old in cultivation.

> "Give me the good old weekday blossoms
> I used to see so long ago,
> With hearty sweetness in their bosoms,
> Ready and glad to bud and blow."

Even were they newcomers, we should speedily care for them, they are so lovable, so winning, so endearing. If I had seen to-day for the first time a Fritillaria, a Violet, a Lilac, a Bluebell, or a Rose, I know it would be a case of love at first sight. But with intimacy they have grown dearer still.

The sense of long-continued acquaintance and friendship which we feel for many garden flowers extends to a few blossoms of field and forest. It is felt to an inexplicable degree by all New Englanders for the Trailing Arbutus, our Mayflower; and it is this unformulated sentiment which makes us like to go to the same spot year after year to gather these beloved flowers. I am sensible of this friendship for Buttercups, they seem the same flowers I knew last year; and I have a distinct sympathy with Owen Meredith's poem: —

> "I pluck the flowers I plucked of old
> About my feet — yet fresh and cold
> The Buttercups do bend;
> The selfsame Buttercups they seem,
> Thick in the bright-eyed green, and such
> As when to me their blissful gleam
> Was all earth's gold — how much!"

We have little of the intense sentiment, the inspiration which filled flower-lovers of olden times. We admire flowers certainly as beautiful works of nature, as objects of wonder in mechanism and in the profusion of growth, and we are occasionally roused to feelings of gratitude to the Maker and Giver of such beauty; but it is not precisely the same regard that the old gardeners and "flowerists" had, which is expressed in this quotation from Gerarde of "the gallant grace of violets": —

> "They admonish and stir up a man to that which is comelie and honest; for flowers through their beautie, varietie of colour and exquisite forme doe bring to a liberall and gentlemanly mind, the remembrance of honestie, comelinesse and all kinds of virtues."

It was a virtue to be comely in those days; as it is indeed a virtue now; and to the pious old herbalists it seemed an impossible thing that any creation which was beautiful should not also be good.

All flowers cannot be loved with equal warmth; it is possible to have a wholesome liking for a flower, a wish to see it around you, which would make you plant it in your borders and treat it well, but which would not be at all akin to love. For others you have a placid tolerance; others you esteem — good, virtuous, worthy creatures, but you cannot warm toward them. Sometimes they have been sung with passion by poets (Swinburne is always glowing over very unresponsive flower souls) and they have been

Meadow Rue.

painted with fervor by artists — and still you do not love them. I do not love Tulips, but I welcome them very cordially in my garden. Others have loved them; the Tulip has had her head turned by attention.

Some flowers we like at first sight, but they do not wear well. This is a hard truth; and I shall not shame the garden-creatures who have done their best to please by betraying them to the world, save in a single case to furnish an example. In late August the Bergamot blossoms in luxuriant heads of white and purplish pink bloom, similar in tint to the abundant Phlox. Both grow freely in the garden of Sylvester Manor. When the Bergamot has romped in your borders for two or three years, you may wish to exile it to a vegetable garden, near the blackberry vines. Is this because it is an herb instead of a purely decorative flower? You never thus thrust out Phlox. A friend confesses to me that she exiled even the splendid scarlet Bergamot after she had grown it for three years in her flower-beds; such subtle influences control our flower-loves.

Beautiful and noble as are the grand contributions of the nineteenth century to us from the garden and fields of Japan and China, we seldom speak of loving them. Thus the Chinese White Wistaria is similar in shape of blossom to the Scotch Laburnum, though a far more elegant, more lavish flower; but the Laburnum is the loved one. I used to read longingly of the Laburnum in volumes of English poetry, especially in Hood's verses, beginning: —

"I remember, I remember,
The house where I was born,"

Ella Partridge had a tall Laburnum tree at her front door; it peeped in the second-story windows. It was so cherished, that I doubt whether its blooms were ever gathered. She told us with conscious pride and rectitude that it was a "yellow Wistaria tree which came from China"; I saw no reason to doubt her words, and as I never chanced to speak to my parents about it, I ever thought of it as a yellow Wistaria tree until I went out into the world and found it was a Scotch Laburnum.

Few garden owners plant now the Snowberry, *Symphoricarpus racemosus*, once seen in every front yard, and even used for hedges. It wasn't a very satisfactory shrub in its habit; the oval leaves were not a cheerful green, and were usually pallid with mildew. The flowers were insignificant, but the clusters of berries were as pure as pearls. In country homes, before the days of cheap winter flowers and omnipresent greenhouses, these snowy clusters were cherished to gather in winter to place on coffins and in hands as white and cold as the berries. Its special offence in our garden was partly on account of this funereal association, but chiefly because we were never permitted to gather its berries to string into necklaces. They were rigidly preserved on the stem as a garden decoration in winter; though they were too closely akin in color to the encircling snowdrifts to be of any value.

In country homes in olden times were found several universal winter posies. On the narrow mantel shelves of farm and village parlors, both in England and America, still is seen a winter posy made of dried stalks of the seed valves of a certain flower; they are shown on the opposite page. Let us see how our old friend, Gerarde, describes this plant: —

> "The stalkes are loden with many flowers like the stocke-gilliflower, of a purple colour, which, being fallen, the seede cometh foorthe conteined in a flat thinne cod, with a sharp point or pricke at one end, in fashion of the moone, and somewhat blackish. This cod is composed of three filmes or skins whereof the two outermost are of an overworne ashe colour, and the innermost, or that in the middle whereon the seed doth hang or cleave, is thin and cleere shining, like a piece of white satten newly cut from the peece."

In the latter clause of this striking description is given the reason for the popular name of the flower, Satin-flower or White Satin, for the inner septum is a shining membrane resembling white satin. Another interesting name is Pricksong-flower. All who have seen sheets of music of Elizabethan days, when the notes of music were called pricks, and the whole sheet a pricksong, will readily trace the resemblance to the seeds of this plant.

Gerarde says it was named "Penny-floure, Money-floure, Silver-plate, Sattin, and among our women called Honestie." The last name was commonly applied at the close of the eighteenth century. It is thus named in writings of Rev. William Hanbury, 1771, and a Boston seedsman then advertised seeds of Honestie "in small quantities, that all might have some." In 1665, Josselyn found White Satin planted and growing plentifully in New England gardens, where I am sure it formed, in garden and house, a happy reminder of their English homes to the wives of the colonists. Since that time it has spread so freely in some localities, especially in southern Connecticut, that it grows wild by the wayside. It is seldom seen now in well-kept gardens, though it should be, for it is really a lovely flower, showing from white to varied and rich light purples. I was charmed with its fresh beauty this spring in the garden of Mrs. Mabel Osgood Wright; a photograph of one of her borders containing Honesty is shown opposite page 174.

Money-in-both-pockets.

At Belvoir Castle in England, in the "Duchess's

Garden," the Satin-flower can be seen in full variety of tint, and fills an important place. It is carefully cultivated by seed and division, all inferior plants being promptly destroyed, while the superior blossoms are cherished.

The flower was much used in charms and spells, as was everything connected with the moon. Drayton's Clarinax sings of Lunaria: —

> "Enchanting lunarie here lies
> In sorceries excelling."

As a child this Lunaria was a favorite flower, for it afforded to us juvenile money. Indeed, it was generally known among us as Money-flower or Money-seed, or sometimes as Money-in-both-pockets. The seed valves formed our medium of exchange and trade, passing as silver dollars.

Through the streets of a New England village there strolled, harmless and happy, one who was known in village parlance as a "softy," one of "God's fools," a poor addle-pated, simple-minded creature, witless — but neither homeless nor friendless; for children cared for him, and feeble-minded though he was, he managed to earn, by rush-seating chairs and weaving coarse baskets, and gathering berries, scant pennies enough to keep him alive; and he slept in a deserted barn, in a field full of rocks and Daisies and Blueberry bushes, — a barn which had been built by one but little more gifted with wits than himself. Poor Elmer never was able to understand that the money which he and the children saved so carefully each autumn from the money plants was not equal in value to the great copper cents of the village store; and when he asked gleefully for a loaf of bread or a quart of

Box Walk in Garden of Frederick J. Kingsbury, Esq. Waterbury, Connecticut.

molasses, was just as apt to offer the shining seed valves in payment as he was to give the coin of the land; and it must be added that his belief received apparent confirmation in the fact that he usually got the bread whether he gave seeds or cents.

He lost his life through his poor simple notion. In the village he was kindly treated by all, clothed, fed, and warmed; but one day there came skulking along the edge of the village what were then rare visitors, two tramps, who by ill-chance met poor Elmer as he was gathering chestnuts. And as the children lingered on their way home from school to take toll of Elmer's store of nuts, they heard him boasting gleefully of his wealth, "hundreds and hundreds of dollars all safe for winter." The children knew what his dollars were, but the tramps did not. Three days of heavy rain passed by, and Elmer did not appear at the store or any house. Then kindly neighbors went to his barn in the distant field, and found him cruelly beaten, with broken ribs and in a high fever, while scattered around him were hundreds of the seeds of his autumnal store of the money plant; these were all the silver dollars his assailants found. He was carried to the almshouse and died in a few weeks, partly from the beating, partly from exposure, but chiefly, I ever believed, from homesickness in his enforced home. His old house has fallen down, but his well still is open, and around it grows a vast expanse of Lunaria, which has spread and grown from the seeds poor Elmer saved, and every year shoots of the tender lilac blooms mingle so charmingly with the white Daisies that the sterile field is one of the show-places of the village, and people drive from afar to see it.

There grow in profusion in our home garden what I always called the Mullein Pink, the Rose Campion (*Lychnis coronaria*). I never heard any one speak

Lunaria in Garden at Waldstein.

of this plant with special affection or admiration; but as a child I loved its crimson flower more than any other flower in the garden. Perhaps I should say I loved the royal color rather than the flower. I gathered tight bunches without foliage into a glowing mass of color unequalled in richness of tint by anything in nature. I have seen only in a stained glass window flooded with high sunlight a crimson approaching that of the Mullein Pink. Gerarde calls the flower the "Gardener's Delight or Gardner's Eie." It was known in French as the Eye of God; and the Rose of Heaven. We used to rub our cheeks with the woolly leaves to give a beautiful rosy blush, and thereby I once skinned one cheek.

Snapdragons were a beloved flower — companions of my childhood in our home garden, but they have been neglected a bit by nearly every one of late years. Plant a clump of the clear yellow and one of pure white Snapdragons, and see how beautiful they are in the garden, and how fresh they keep when cut. We had such a satisfying bunch of them on the dinner table to-day, in a milk-white glazed Chinese jar; yellow Snapdragons, with "borrowed leaves" of Virgin's-bower (*Adlumia*) and a haze of Gypsophila over all.

A flower much admired in gardens during the early years of the nineteenth century was the Plume Poppy (*Bocconia*). It has a pretty pinkish bloom in general shape somewhat like Meadow Rue (see page 164 and page 167). A friend fancied a light feathery look over certain of her garden borders, and she planted plentifully Plume Poppy and Meadow Rue; this was in 1895. In 1896 the effect was exquisite; in 1897 the garden feathered out with far too much fulness; in 1901 all the combined forces of all the weeds of the garden could not equal these two flowers in utter usurpment and close occupation of every inch of that garden. The Plume Poppy has a strong tap-root which would be a good symbol of the root of the tree Ygdrassyl — the Tree of Life, that never dies. You can go over the borders with scythe and spade and hoe, and even with manicure-scissors, but roots of the Plume Poppy will still hide and send up vigorous growth the succeeding year.

We have grown so familiar with some old doubled blossoms that we think little of their being double. One such, symmetrical of growth, beautiful of foliage, and gratifying of bloom, is the Double Buttercup. It is to me distinctly one of our most old-fashioned flowers in aspect. A hardy great clump of many years' growth is one of the ancient treasures of our garden; its golden globes are known in England as Bachelor's Buttons, and are believed by many to be the Bachelor's Buttons of Shakespeare's day.

Dahlias afford a striking example of the beauty of single flowers when compared to their doubled descendants. Single Dahlias are fine flowers, the yellow and scarlet ones especially so. I never thought double Dahlias really worth the trouble spent on them in our Northern gardens; so much staking and tying, and fussing, and usually an autumn storm wrenches them round and breaks the stem or a frost nips them just as they are in bloom. A Dahlia hedge or a walk such as this one at Ravens-

Dahlia Walk at Ravensworth.

worth, Virginia, is most stately and satisfying. I like, in moderation, many of the smaller single and double Sunflowers. Under the reign of *Patience*, the Sunflower had a fleeting day of popularity, and flaunted in garden and parlor. Its place was false. It was never a garden flower in olden times, in the sense of being a flower of ornament or beauty; its place was in the kitchen garden, where it belongs.

Peas have ever been favorites in English gardens since they were brought to England. We have all seen the print, if not the portrait, of Queen Elizabeth garbed in a white satin robe magnificently embroidered with open pea-pods and butterflies. A "City of London Madam" had a delightful head ornament of open pea-pods filled with peas of pearls; this was worn over a hood of gold-embroidered muslin, and with dyed red hair, must have been a most modish affair. Sweet Peas have had a unique history. They have been for a century a much-loved flower of the people both in England and America, and they were at home in cottage borders and fine gardens; were placed in vases, and carried in nosegays and posies; were loved of poets — Keats wrote an exquisite characterization of them. They had beauty of color, and a universally loved perfume — but florists have been blind to them till within a few years. A bicentenary exhibition of Sweet Peas was given in London in July, 1900; now there is formed a Sweet Pea Society. But no societies and no exhibitions ever will make them a "florist's flower"; they are of value only for cutting; their habit of growth renders them useless as a garden decoration.

We all take notions in regard to flowers, just as we do in regard to people. I hear one friend say,

"I love every flower that grows," but I answer with emphasis, "I don't!" I have ever disliked the Portulaca,—I hate its stems. It is my fate never to escape it. I planted it once to grow under Sweet Alyssum in the little enclosure of earth behind my city home; when I returned in the autumn, everything was covered, blanketed, overwhelmed with Portulaca. Since then it comes up even in the grass, and seems to thrive by being trampled upon. The Portulaca was not a flower of colonial days; I am glad to learn our great-grandmothers were not pestered with it; it was not described in the *Botanical Magazine* till 1829.

I do not care for the Petunia close at hand on account of its sickish odor. But in the dusky border the flowers shine like white stars (page 180), and make you almost forgive their poor colors in the daylight. I never liked the Calceolaria. Every child in our town used to have a Calceolaria in her own small garden plot, but I never wanted one. I care little for Chrysanthemums; they fill in the border in autumn, and they look pretty well growing, but I like few of the flowers close at hand. By some curious twist of a brain which, alas! is apt not to deal as it is expected and ought to, with sensations furnished to it, I have felt this distaste for Chrysanthemums since I attended a Chrysanthemum Show. Of course, I ought to love them far more, and have more eager interest in them—but I do not. Their sister, the China Aster, I care little for. The Germans call Asters "death-flowers." The Empress of Austria at the Swiss hotel where she lodged just before she was murdered, found the rooms decorated with China Asters. She said to her attendant that the flowers were in Austria termed death-flowers — and so they proved. The Aster is among the flowers prohibited in Japan for felicitous occasions, as are the Balsam, Rhododendron, and Azalea.

Petunias.

Those who read these pages may note perhaps that I say little of Lilies. I do not care as much for them as most garden lovers do. I like all our wild Lilies, especially the yellow Nodding Lily of our fields; and the Lemon Lily of our gardens is ever a delight; but the stately Lilies which are such general favorites, Madonna Lilies, Japan Lilies, the Gold-banded Lilies, are not especially dear to me.

I love climbing vines, whether of delicate leaf or beautiful flower. In a room I place all the decoration that I can on the walls, out of the way, leaving thus space to move around without fear of displacement or injury of fragile things; so in a limited garden space, grass room under our feet, with flowering vines on the surrounding walls are better than many crowded flower borders. A tiny space can quickly be made delightful with climbing plants. The common Morning-glory, called in England the Bell-bind, is frequently advertised by florists of more encouragement than judgment, as suitable to plant freely in order to cover fences and poor sandy patches of ground with speedy and abundant leafage and bloom. There is no doubt that the Morning-glory will do all this and far more than is promised. It will also spread above and below ground from the poor strip of earth to every other corner of garden and farm. This it has done till, in our Eastern states, it is now classed as a wild flower. It will never look wild, however, meet it where you will. It is as domestic and tame as a barnyard fowl, which, wandering in the wildest woodland, could never be mistaken as game. The garden at Claymont, the Virginia home of Mr. Frank R. Stockton, afforded a striking example of the spreading and strangling properties of the Morning-glory, not under encouragement, but simply under toleration. Mr. Stockton tells me that the entire expanse of his yards and garden, when he first saw them, was a solid mass of Morning-glory blooms. Every stick, every stem, every stalk, every shrub and blade of grass, every vegetable growth, whether dead or alive, had its encircling and overwhelming Morning-glory companion, set full of tiny undersized blossoms of varied tints. It was a beautiful sight at break of day, — a vast expanse of acres jewelled with Morning-glories — but it wasn't the new owner's notion of a flower garden.

In my childhood flower agents used to canvass country towns from house to house. Sometimes they had a general catalogue, and sold many plants, trees, and shrubs. Oftener they had but a single plant which they were "booming." I suspect that their trade came through the sudden introduction of so many and varied flowers and shrubs from China and Japan. I am told that the first Chinese Wistarias and a certain Fringe tree were sold in this manner; and I know the white Hydrangea was, for I recall it, though I do not know that this was its first sale. I remember too that suddenly half the houses in town, on piazza or trellis, had the rich purple blooms of the *Clematis Jackmanni;* for a very persuasive agent had gone through the town the previous year. Of course people of means bought then, as now, at nurseries; but at many humble homes, whose owners would never have thought of buying from a greenhouse, he sold his plants. It gave an agreeable rivalry, when all started plants together, to see whose flourished best and had the amplest bloom. Thoreau recalled the pleasant emulation of many owners in Concord of a certain Rhododendron, sold thus sweepingly by an agent. The purple Clematis displaced an old climbing favorite, the Trumpet Honeysuckle, once seen by every door. It was so beloved of humming-birds and so beautiful, I wonder we could ever destroy it. Its downfall was hastened by its being infested by a myriad of tiny green aphides, which proceeded from it to our Roses. I recall well these little plant insects, for I was very fond of picking the tubes of the Honeysuckle for the drop of pure honey within, and I had to abandon reluctantly the sweet morsels.

We have in our garden, and it is shown on the succeeding page, a vine which we carefully cherished in seedlings from year to year, and took much pride in. It came to us with the Ambrosia from the Walpole garden. It was not common in gardens in our neighborhood, and I always looked upon it as something very choice, and even rare, as it certainly was something very dainty and pretty. We called it Virgin's-bower. When I went out into the world I found that it was not rare, that it grew wild from Connecticut to the far West; that it was Climbing Fumitory, or Mountain Fringe, *Adlumia*. When Mrs. Margaret Deland asked if we had Alleghany Vine in our garden, I told her I had never seen it, when all the while it was our own dear Virgin's-bower. It doesn't seem hardy enough to be a wild thing; how could it make its way against the fierce vines and thorns of the forest when it hasn't a bit of woodiness in its stems and its leaves and flowers are so tender! I cannot think any garden perfect without it, no matter what else is there, for its delicate green Rue-like leaves lie so gracefully on stone or brick walls, or on fences, and it trails its slender tendrils so lightly over dull shrubs that are out of flower, beautifying them afresh with an alien bloom of delicate little pinkish blossoms like tiny Bleeding-hearts.

Another old favorite was the Balloon-vine, some-

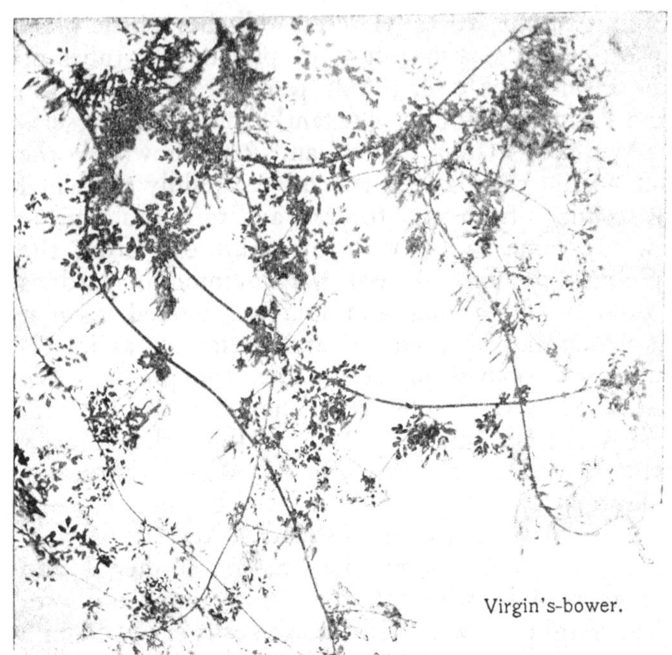
Virgin's-bower.

times called Heartseed or Heart-pea, with its seeds like fat black hearts, with three lobes which made them globose instead of flat. This, too, had pretty compound leaves, and the whole vine, like our Virgin's-bower, lay lightly on what it covered; but the Dutchman's-pipe had a leafage too heavy save to make a thick screen or arch quickly and solidly. It did well enough in gardens which had not had a long cultivated past, or made little preparation for a cherished future; but it certainly was not suited to our garden, where things were not planted for a day. These three are native vines of rich woods in our Central and Western states. The Matrimony-vine was an old favorite; one from the porch of the Van Cortlandt manor-house, over a hundred years old, is shown on the next page. Often you see a straggling, sprawling growth; but this one is as fine as any vine could be.

Patient folk — as were certainly those of the old-time gardens, tried to keep the Rose Acacia as a favorite. It was hardy enough, but so hopelessly brittle in wood that it was constantly broken by the wind and snow of our Northern winters, even though it was sheltered under some stronger shrub. At the end of a lovely Salem garden, I beheld this June a long row of Rose Acacias in full bloom. I am glad I possess in my memory the exquisite harmony of their shimmering green foliage and rosy flower clusters. Miss Jekyll, ever resourceful, trains the Rose Acacia on a wall; and fastens it down by planting sturdy Crimson Ramblers by its side; her skilful example may well be followed in America and thus restore to our gardens this beautiful flower.

One flower, termed old-fashioned by nearly every one, is really a recent settler of our gardens. A popular historical novel of American life at the time of the Revolution makes the hero and heroine play a very pretty love scene over a spray of the Bleeding-heart, the Dielytra, or Dicentra. Unfortunately for the truth of the novelist's picture, the Dielytra was not introduced to the gardens of English-speaking folk till 1846, when the London Horticultural Society received a single plant from the north of China.

Matrimony-vine at Van Cortlandt Manor.

How quickly it became cheap and abundant; soon it bloomed in every cottage garden; how quickly it became beloved! The graceful racemes of pendant rosy flowers were eagerly welcomed by children; they have some inexplicable, witching charm ; even young children in arms will stretch out their little hands and attempt to grasp the Dielytra, when showier blossoms are passed unheeded. Many tiny playthings can be formed of the blossoms : only deft fingers can shape the delicate lyre in the "frame." One of its folk names is "Lyre flower"; the two wings can be bent back to form a gondola.

We speak of modern flowers, meaning those which have recently found their way to our gardens. Some of these clash with the older occupants, but one has promptly been given an honored place, and appears so allied to the older flowers in form and spirit that it seems to belong by their side — the *Anemone Japonica.* Its purity and beauty make it one of the delights of the autumn garden; our grandmothers would have rejoiced in it, and have divided the plants with each other till all had a row of it in the garden borders. In its red form it was first pictured in the *Botanical Magazine*, in 1847, but it has been commonly seen in our gardens for only twenty or thirty years.

These two flowers, the *Dielytra spectabilis* and *Anemone Japonica*, are among the valuable gifts which our gardens received through the visits to China of that adventurous collector, Robert Fortune. He went there first in 1842, and for some years constantly sent home fresh treasures. Among the best-known garden flowers of his introducing are the two named above, and *Kerria Japonica*, *Forsythia viridissima*, *Weigela rosea*, *Gardenia Fortunāna*, *Daphne Fortunei*, *Berberis Fortunei*, *Jasminum nudiflorum*, and many varieties of Prunus, Viburnum, Spiræa, Azalea, and Chrysanthemum. The fine yellow Rose known as Fortune's Yellow was acquired by him during a venturesome trip which he took, disguised as a Chinaman. The white Chinese Wistaria is regarded as the most important of his collections. It is deemed by some

White Wistaria.

flower-lovers the most exquisite flower in the entire world. The Chinese variety is distinguished by the length of its racemes, sometimes three feet long. The lower part of a vine of unusual luxuriance and beauty is shown above This special vine flowers in full richness of bloom every alternate year, and this photograph was taken during its "poor year"; for in its finest inflorescence its photograph would show simply a mass of indistinguishable whiteness. Mr. Howell has named it The Fountain, and above the pouring of white blossoms shown in this picture is an upper cascade of bloom. This Wistaria is not growing in an over-favorable locality, for winter winds are bleak on the southern shores of Long Island; but I know no rival of its beauty in far warmer and more sheltered sites.

Many of the Deutzias and Spiræas which beautify our spring gardens were introduced from Japan before Fortune's day by Thunberg, the great exploiter of Japanese shrubs, who died in 1828. The Spiræa Van Houtteii (facing page 190) is perhaps the most beautiful of all. Dean Hole names the Spiræas, Deutzias, Weigelas, and Forsythias as having been brought into his ken in English gardens within his own lifetime, that is within fourscore years.

In New England gardens the Forsythia is called 'Sunshine Bush' — and never was folk name better bestowed, or rather evolved. For in the eager longing for spring which comes in the bitterness of March, when we cry out with the poet, "O God, for one clear day, a Snowdrop and sweet air," in our welcome to fresh life, whether shown in starting leaf or frail blossom, the Forsythia shines out a grateful delight to the eyes and heart, concentrating for a week all the golden radiance of sunlight, which later will be shared by sister shrubs and flowers. *Forsythia suspensa*, falling in long sweeps of yellow bells, is in some favorable places a cascade of liquid light. No shrub in our gardens is more frequently ruined by gardeners than these Forsythias. It takes an artist to prune the *Forsythia suspensa.* You can steal the sunshine for your homes ere winter is gone by breaking long sprays of the Sunshine Bush and placing them in tall deep jars of water. Split up the ends of the stems that they may absorb plentiful water, and the golden plumes will soon open to fullest glory within doors.

There is another yellow flowered shrub, the Corchorus, which seems as old as the Lilac, for it is ever found in old gardens; but it proves to be a Japanese shrub which we have had only a hundred years. The little, deep yellow, globular blossoms appear in early spring and sparsely throughout the whole summer. The plant isn't very adorning in its usual ragged growth, but it was universally planted.

It may be seen from the shrubs of popular growth which I have named that the present glory of our shrubberies is from the Japanese and Chinese shrubs, which came to us in the nineteenth century through Thunberg, Fortune, and other bold collectors. We had no shrub-sellers of importance in the eighteenth century; the garden lover turned wholly to the seedsman and bulb-grower for garden supplies, just as we do to-day to fill our old-fashioned gardens. The new shrubs and plants from China and Japan did not clash with the old garden flowers, they seemed like kinsfolk who had long been separated and rejoiced in being reunited; they were indeed fellow-countrymen. We owed scores of our older flowers to the Orient, among them such important ones as the Lilac, Rose, Lily, Tulip, Crown Imperial.

Spiræa Van Houtteii.

We can fancy how delighted all these Oriental shrubs and flowers were to meet after so many years of separation. What pleasant greetings all the cousins must have given each other; I am sure the Wistaria was glad to see the Lilac, and the Fortune's Yellow Rose was duly respectful to his old cousin, the thorny yellow Scotch Rose. And I seem to hear a bit of scandal passing from plant to plant! Listen! it is the Bleeding-heart gossiping with the Japanese Anemone: "Well! I never thought that Lilac girl would grow to be such a beauty. So much color! Do you suppose it can be natural? Mrs. Tulip hinted to me yesterday that the girl used fertilizers, and it certainly looks so. But she can't say much herself — I never saw such a change in any creatures as in those Tulips. You remember how commonplace their clothes were? Now such extravagance! Scores of gowns, and all made abroad, and at *her* age! Here are you and I, my dear, both young, and we really ought to have more clothes. I haven't a thing but this pink gown to put on. It's lucky you had a white gown, for no one liked your pink one. Here comes Mrs. Rose! How those Rose children have grown! I never should have known them."

CHAPTER VIII

COMFORT ME WITH APPLES

"What can your eye desire to see, your eares to heare, your mouth to taste, or your nose to smell, that is not to be had in an Orchard? with Abundance and Variety? What shall I say? 1000 of Delights are in an Orchard; and sooner shall I be weary than I can reckon the least part of that pleasure which one, that hath and loves an Orchard, may find therein."

—*A New Orchard*, WILLIAM LAWSON, 1618.

N every old-time garden, save the revered front yard, the borders stretched into the domain of the Currant and Gooseberry bushes, and into the orchard. Often a row of Crabapple trees pressed up into the garden's precincts and shaded the Sweet Peas. Orchard and garden could scarcely be separated, so closely did they grow up together. Every old garden book had long chapters on orchards, written *con amore*, with a zest sometimes lacking on other pages. How they loved in the days of Queen Elizabeth and of Queen Anne to sit in an orchard, planted, as Sir Philip Sidney said, "cunningly with trees of taste-pleasing fruits." How charming were their orchard seats, "fachoned for meditacon!" Sometimes these orchard seats were banks of the strongly scented Camomile, a favorite plant of Lord Bacon's day. Wordsworth wrote in jingling rhyme:—

> "Beneath these fruit-tree boughs that shed
> Their snow-white blossoms on my head,
> With brightest sunshine round me spread
> Of spring's unclouded weather,
> In this sequester'd nook how sweet
> To sit upon my orchard seat;
> And flowers and birds once more to greet,
> My last year's friends together."

The incomparable beauty of the Apple tree in full bloom has ever been sung by the poets, but even their words cannot fitly nor fully tell the delight to the senses of the close view of those exquisite pink and white domes, with their lovely opalescent tints, their ethereal fragrance; their beauty infinitely surpasses that of the vaunted Cherry plantations of Japan. In the hand the flowers show a distinct ruddiness, a promise of future red cheeks; but a long vista of trees in bloom displays no tint of pink, the flowers seem purest white. Looking last May across the orchard at Hillside, adown the valley of the Hudson with its succession of blossoming orchards, we could paraphrase the words of Longfellow's *Golden Legend:*—

> "The valley stretching below
> Is white with blossoming Apple trees, as if touched with lightest snow."

In the darkest night flowering Apple trees shine with clear radiance, and an orchard of eight hundred acres, such as may be seen in Niagara County, New York, shows a white expanse like a lake of quicksilver. This county, and its neighbor, Orleans County, form an Apple paradise — with their orchards of fifty and even a hundred thousand trees.

Apple Trees at White Hall, the Home of Bishop Berkeley.

The largest Apple tree in New England is in Cheshire, Connecticut. Its trunk measures, one foot above all root enlargements, thirteen feet eight inches in circumference.

Its age is traced back a hundred and fifty years.

At White Hall, the old home of Bishop Berkeley in the island of Rhode Island, still stand the Apple trees of his day. A picture of them is shown on page 194.

The sedate and comfortable motherliness of old Apple trees is felt by all Apple lovers. John Burroughs speaks of "maternal old Apple trees, regular old grandmothers, who have seen trouble." James Lane Allen, amid his apostrophes to the Hemp plant, has given us some beautiful glimpses of Apple trees and his love for them. He tells of "provident old tree mothers on the orchard slope, whose red-cheeked children are autumn Apples." It is this motherliness, this domesticity, this homeliness that makes the Apple tree so cherished, so beloved. No scene of life in the country ever seems to me homelike if it lacks an Apple orchard — this doubtless, because in my birthplace in New England they form a part of every farm scene, of every country home. Apple trees soften and humanize the wildest country scene. Even in a remote pasture, or on a mountain side, they convey a sentiment of home; and after being lost in the mazes of close-grown wood-roads Apple trees are inexpressibly welcome as giving promise of a sheltering roof-tree. Thoreau wrote of wild Apples, but to me no Apples ever look wild. They may be the veriest Crabs, growing in wild spots, unbidden, and savage and bitter in their tang, but even these seedling Pippins are domestic in aspect.

On the southern shores of Long Island, where meadow, pasture, and farm are in soil and crops like New England, the frequent absence of Apple orchards makes these farm scenes unsatisfying, not homelike. No other fruit trees can take their place. An Orange tree, with its rich glossy foliage, its perfumed ivory flowers and buds, and abundant golden fruit, is an exquisite creation of nature; but an Orange grove has no ideality. All fruit trees have a beautiful inflorescence — few have sentiment. The tint of a blossoming Peach tree is perfect; but I care not for a Peach orchard. Plantations of healthy Cherry trees are lovely in flower and fruit time, whether in Japan or Massachusetts, and a Cherry tree is full of happy child memories; but their tree forms in America are often disfigured with that ugly fungous blight which is all the more disagreeable to us since we hear now of its close kinship to disease germs in the animal world.

I cannot see how they avoid having Apple trees on these Long Island farms, for the Apple is fully determined to stand beside every home and in every garden in the land. It does not have to be invited; it will plant and maintain itself. Nearly all fruits and vegetables which we prize, depend on our planting and care, but the Apple is as independent as the New England farmer. In truth Apple trees would grow on these farms if they were loved or even tolerated, for I find them forced into Long Island hedge-rows as relentlessly as are forest trees.

The Indians called the Plantain the "white man's foot," for it sprung up wherever he trod; the Apple tree might be called the white man's shadow. It is the Vine and Fig tree of the temperate zone, and might be chosen as the totem of the white settlers. Our love for the Apple is natural, for it was the characteristic fruit of Britain; the clergy were its chief cultivators; they grew Apples in their monastery gardens, prayed for them in special religious ceremonies, sheltered the fruit by laws, and even

"The valley stretching below
Is white with blossoming Apple trees, as if touched with lightest snow."

named the Apple when pronouncing the blessings of God upon their princes and rulers.

Thoreau described an era of luxury as one in which men cultivate the Apple and the amenities of the garden. He thought it indicated relaxed nerves to read gardening books, and he regarded gardening as a civil and social function, not a love of nature. He tells of his own love for freedom and savagery — and he found what he so deemed at

Walden Pond. I am told his haunts are little changed since the years when he lived there; and I had expected to find Walden Pond a scene of much wild beauty, but it was the mildest of wild

Old Hand-power Cider Mill.

woods; it seemed to me as thoroughly civilized and social as an Apple orchard.

Thoreau christened the Apple trees of his acquaintance with appropriate names in the *lingua vernacula:* the Truant's Apple, the Saunterer's Apple, December Eating, Wine of New England, the Apple of the Dell in the Wood, the Apple of the Hollow in the Pasture, the Railroad Apple, the Cellar-hole Apple, the Frozen-thawed, and many more; these he loved for their fruit; to them let me add the Playhouse Apple trees, loved solely for their ingeniously twisted branches, an Apple tree of the garden, often overhanging the flower borders. I recall their glorious whiteness in the spring, but I cannot remember that they bore any fruit save a group of serious little girls. I know there were no Apples on the Playhouse Apple trees in my garden, nor on the one in Nelly Gilbert's or Ella Partridge's garden. There is no play place for girls like an old Apple tree. The main limbs leave the trunk at exactly the right height for children to reach, and every branch and twig seems to grow and turn only to form delightful perches for children to climb among and cling to. Some Apple trees in our town had a copy of an Elizabethan garden furnishing; their branches enclosed tree platforms about twelve feet from the ground, reached by a narrow ladder or flight of steps. These were built by generous parents for their children's playhouses, but their approach of ladder was too unhazardous, their railings too safety-assuring, to prove anything but conventional and uninteresting. The natural Apple tree offered infinite variety, and a slight sense of daring to the climber. Its possibility of accident was fulfilled; untold number of broken arms and ribs — juvenile — were resultant from falls from Apple trees.

One of Thoreau's Apples was the Green Apple (*Malus viridis*, or *Cholera morbifera puerelis delectissima*). I know not for how many centuries boys (and girls too) have eaten and suffered from green apples. A description was written in 1684 which

Pressing out Cider in Old Hand Mill.

might have happened any summer since; I quote it with reminiscent delight, for I have the same love for the spirited relation that I had in my early youth when I never, for a moment, in spite of the significant names, deemed the entire book anything but a real story; the notion that *Pilgrim's Progress* was an allegory never entered my mind.

> "Now there was on the other side of the wall a *Garden*. And some of the Fruit-Trees that grew in the Garden shot their Branches over the Wall, and being mellow, they that found them did gather them up and oft eat of them to their hurt. So *Christiana's* Boys, *as Boys are apt to do*, being *pleas'd* with the Trees did *Plash* them and began to eat. Their Mother did also chide them for so doing, but still the Boys went on. Now *Matthew* the Eldest Son of *Christiana* fell sick. . . . There dwelt not far from thence one Mr. *Skill* an Antient and well approved Physician. So Christiana desired it and they sent for him and he came. And when he was entered the Room and a little observed the Boy he concluded that he was sick of the Gripes. Then he said to his Mother, *What Diet has Matthew of late fed upon? Diet*, said Christiana, *nothing but which is wholesome.* The Physician answered, *This Boy has been tampering with something that lies in his Maw undigested. . . .* Then said Samuel, *Mother, Mother, what was that which my brother did gather up and eat. You know there was an Orchard and my Brother did plash and eat. True, my child*, said Christiana, *naughty boy as he was. I did chide him and yet he would eat thereof.*"

The realistic treatment of Mr. Skill and Matthew's recovery thereby need not be quoted.

An historic Apple much esteemed in Connecticut and Rhode Island, and often planted at the edge of the flower garden, is called the Sapson, or Early Sapson, Sapson Sweet, Sapsyvine, and in Pennsylvania, Wine-sap. The name is a corruption of the old English Apple name, Sops-o'-wine. It is a charming little red-cheeked Apple of early autumn, slightly larger than a healthy Crab-apple. The clear red of its skin perfuses in coral-colored veins and beautiful shadings to its very core. It has a condensed, spicy, aromatic flavor, not sharp like a Crab-apple, but it makes a better jelly even than the Crab-apple — jelly of a ruby color with an almost wine-like flavor, a true Sops-of-wine. This fruit is deemed so choice that I have known the sale of a farm to halt for some weeks until it could be proved that certain Apple trees in the orchard bore the esteemed Sapsyvines.

Under New England and New York farm-houses was a cellar filled with bins for vegetables and apples. As the winter passed on there rose from these cellars a curious, earthy, appley smell, which always seemed most powerful in the best parlor, the room least used. How Schiller, who loved the scent of rotten apples, would have rejoiced! The cellar also contained many barrels of cider; for the beauty of the Apple trees, and the use of their fruit as food, were not the only factors which influenced the planting of the many Apple orchards of the new world; they afforded a universal drink — cider. I have written at length, in my books, *Home Life in Colonial Days* and *Stage-Coach and Tavern Days*, the history of the vogue and manufacture of cider in the new world. The cherished Apple orchards of Endicott, Blackstone, Wolcott, and Winthrop were so speedily multiplied that by 1670 cider was plentiful and cheap everywhere. By the opening of the eighteenth century it had wholly crowded out beer and metheglin; and was the drink of old and young on all occasions.

At first, cider was made by pounding the Apples by hand in wooden mortars; then simple mills were formed of a hollowed log and a spring board. Rude hand presses, such as are shown on pages 198 and 200, were known in 1660, and lingered to our

Old Horse-lever Cider Mill.

own day. Kalm, the Swedish naturalist, saw ancient horse presses (like the one depicted on this page) in use in the Hudson River Valley in 1749. In autumn the whole country-side was scented with the sour, fruity smell from these cider mills; and the gift of a draught of sweet cider to any passer-by was as ample and free as of water from the brook-side. The cider when barrelled and stored for winter was equally free to all comers, as well it might be, when many families stored a hundred barrels for winter use.

The Washingtonian or Temperance reform which swept over this country like a purifying wind in the

"Straining off" the Cider.

first quarter of the nineteenth century, found many temporizers who tried to exclude cider from the list of intoxicating drinks which converts pledged themselves to abandon. Some farmers who adopted this much-needed movement against the all-prevailing vice of drunkenness received it with fanatic zeal. It makes the heart of the Apple lover ache to read that in this spirit they cut down whole orchards of flourishing Apple trees, since they could conceive no adequate use for their apples save for cider. That any should have tried to exclude cider from the list of intoxicating beverages seems barefaced indeed to those who have tasted that most potent of all spirits — frozen cider. I once drank a small modicum of Jericho cider, as smooth as Benedictine and more persuasive, which made a raw day in April seem like sunny midsummer. I afterward learned from the ingenuous Long Island farmer whose hospitality gave me this liqueur that it had been frozen seven times. Each time he had thrust a red-hot poker into the bung-hole of the barrel, melted all the watery ice and poured it out; therefore the very essence of the cider was all that remained.

It is interesting to note the folk customs of Old England which have lingered here, such as domestic love divinations. The poet Gay wrote: —

"I pare this Pippin round and round again,
My shepherd's name to flourish on the plain.
I fling th' unbroken paring o'er my head,
Upon the grass a perfect L. is read."

I have seen New England schoolgirls, scores of times, thus toss an "unbroken paring." An ancient trial of my youth was done with Apple seeds; these were named for various swains, then slightly wetted and stuck on the cheek or forehead, while we chanted: —

"Pippin! Pippin! Paradise!
Tell me where my true love lies!"

The seed that remained longest in place indicated the favored and favoring lover.

With the neglect in this country of Saints' Days and the Puritanical frowning down of all folk customs connected with them, we lost the delightful wassailing of the Apple trees. This, like many another religious observance, was a relic of heathen sacrifice, in this case to Pomona. It was celebrated with slight variations in various parts of England; and was called an Apple howling, a wassailing, a youling, and other terms. The farmer and his workmen carried to the orchard great jugs of cider or milk pans filled with cider and roasted apples. Encircling in turn the best bearing trees, they drank from "clayen-cups," and poured part of the contents on the ground under the trees. And while they wassailed the trees they sang: —

"Here's to thee, old Apple tree!
Whence thou mayst bud, and whence thou mayst blow,
And whence thou mayst bear Apples enow!
Hats full! caps full,
Bushel — Bushel — sacks full,
And my pockets full too."

Another Devonshire rhyme ran: —

"Health to thee, good Apple tree!
Well to bear pocket-fulls, hat-fulls,
Peck-fulls, bushel bag-fulls."

The wassailing of the trees gave place in America to a jovial autumnal gathering known as an Apple cut, an Apple paring, or an Apple bee. The cheerful kitchen of the farm-house was set out with its entire array of empty pans, pails, tubs, and baskets. Heaped-up barrels of apples stood in the centre of the room. The many skilful hands of willing neighbors emptied the barrels, and with sharp knives or an occasional Apple parer, filled the empty vessels with cleanly pared and quartered apples.

When the work was finished, divinations with Apple parings and Apple seeds were tried, simple country games were played; occasionally there was a fiddler and a dance. An autumnal supper was served from the three zones of the farm-house: nuts from the attic, Apples from the pantry, and cider from the cellar. The apple-quarters intended for drying were strung on homespun linen thread and hung out of doors on clear drying days. A humble hillside home in New Hampshire thus quaintly festooned is shown in the illustration opposite page 208 — a characteristic New Hampshire landscape. When thoroughly dried in sun and wind, these sliced apples were stored for the winter by being hung from rafter to rafter of various living rooms, and remained thus for months (gathering vast accumulations of dust and germs for our blissfully ignorant and unsqueamish grandparents) until the early days of spring, when Apple sauce, Apple butter, and the stores of Apple bin and Apple pit were exhausted, and they then afforded, after proper baths and soakings, the wherewithal for that domestic comestible — dried Apple pie. The Swedish parson, Dr. Acrelius, writing home to Sweden in

1758 an account of the settlement of Delaware, said : —

" Apple pie is used throughout the whole year, and when fresh Apples are no longer to be had, dried ones are used. It is the evening meal of children. House pie, in country places, is made of Apples neither peeled nor freed from their cores, and its crust is not broken if a wagon wheel goes over it."

I always had an undue estimation of Apple pie in my childhood, from an accidental cause : we were requested by the conscientious teacher in our Sunday-school to " take out " each week without fail from the " Select Library " of the school a " Sabbath-school Library Book." The colorless, albeit pious, contents of the books classed under that title are well known to those of my generation; even such a child of the Puritans as I was could not read them. There were two anchors in that sea of despair, — but feeble holds would they seem to-day, — the first volumes of *Queechy* and *The Wide, Wide World.* With the disingenuousness of childhood I satisfied the rules of the school and my own conscience by carrying home these two books, and no others, on alternate Sundays for certainly two years. The only wonder in the matter was that the transaction escaped my Mother's eye for so long a time. I read only isolated scenes; of these the favorite was the one wherein Fleda carries to the woods for the hungry visitor, who was of the English nobility, several large and toothsome sections of green Apple pie and cheese. The prominence given to that Ap-

Drying Apples.

ple pie in that book and in my two years of reading idealized it. On a glorious day last October I drove to New Canaan, the town which was the prototype of Queechy. Hungry as ever in childhood from the clear autumnal air and the long drive from Lenox, we asked for luncheon at what was reported to be a village hostelry. The exact counterpart of Miss Cynthia Gall responded rather sourly that she wasn't "boarding or baiting" that year. Humble entreaties for provender of any kind elicited from her for each of us a slice of cheese and a large and truly noble section of Apple pie, the very pie of Fleda's tale, which we ate with a bewildered sense as of a previous existence. This was intensified as we strolled to the brook under the Queechy Sugar Maples, and gathered there the great-grandchildren of Fleda's Watercresses, and heard the sound of Hugh's sawmills.

Six hundred years ago English gentlewomen and goodwives were cooking Apples just as we cook them now — they even had Apple pie. A delightful recipe of the fourteenth century was for "Appeluns for a Lorde, in opyntide." Opyntide was springtime; this was, therefore, a spring dish fit for a lord.

Apple-moy and Apple-mos, Apple Tansy, and Pommys-morle were delightful dishes and very rich food as well. The word pomatum has now no association with *pomum*, but originally pomatum was made partly of Apples. In an old "Dialog between Soarness and Chirurgi," written by one Dr. Bulleyne in the days of Queen Elizabeth, is found this question and its answer: —

"*Soarness.* How make you pomatum?

"*Chirurgi.* Take the fat of a yearly kyd one pound, temper it with the water of musk-roses by the space of foure dayes, then take five apples, and dresse them, and cut them in pieces, and lard them with cloves, then boyl them altogeather in the same water of roses in one vessel of glasse set within another vessel, let it boyl on the fyre so long tyll it all be white, then wash them with the same water of muske-roses, this done kepe it in a glasse and if you will have it to smell better, then you must put in a little civet or musk, or both, or ambergrice. Gentil women doe use this to make theyr faces fayr and smooth, for it healeth cliftes in the lippes, or in any places of the hands and face."

With the omission of the civet or musk I am sure this would make to-day a delightful cream; but there is one condition which the "gentil woman" of to-day could scarcely furnish — the infinite patience and leisure which accompanied and perfected all such domestic work three centuries ago. A pomander was made of "the maste of a sweet Apple tree being gathered betwixt two Lady days," mixed with various sweet-scented drugs and gums and Rose leaves, and shaped into a ball or bracelet.

The successor of the pomander was the Clove Apple, or "Comfort Apple," an Apple stuck solidly with cloves. In country communities, one was given as an expression of sympathy in trouble or sorrow. Visiting a country "poorhouse" recently, we were shown a "Comfort-apple" which had been sent to one of the inmates by a friend; for even paupers have friends.

"Taffaty tarts" were of paste filled with Apples sweetened and seasoned with Lemon, Rose-water, and Fennel seed. Apple-sticklin', Apple-stucklin, Apple-twelin, Apple-hoglin, are old English provincial names of Apple pie; Apple-betty is a New

Ancient Apple Picker, Apple Racks, Apple Parers, Apple-butter Kettle, Apple-butter Paddle, Apple-butter Stirrer, Apple-butter Crocks.

England term. The Apple Slump of New England homes was not the "slump-pye" of old England, which was a rich mutton pie flavored with wine and jelly, and covered with a rich confection of nuts and fruit.

In Pennsylvania, among the people known as the Pennsylvania Dutch, the Apple frolic was universal.

Each neighbor brought his or her own Apple parer. This people make great use of Apples and cider in their food, and have many curious modes of cooking them. Dr. Heilman in his paper on "The Old Cider Mill" tells of their delicacy of "cider time" called cider soup, made of equal parts of cider and water, boiled and thickened with sweet cream and flour; when ready to serve, bits of bread or toast are placed in it. "Mole cider" is made of boiling cider thickened to a syrup with beaten eggs and milk. But of greatest importance, both for home consumption and for the market, is the staple known as Apple butter. This is made from sweet cider boiled down to about one-third its original quantity. To this is added an equal weight of sliced Apples, about a third as much of molasses, and various spices, such as cloves, ginger, mace, cinnamon or even pepper, all boiled together for twelve or fifteen hours. Often the great kettle is filled with cider in the morning, and boiled and stirred constantly all day, then the sliced Apples are added at night, and the monotonous stirring continues till morning, when the butter can be packed in jars and kegs for winter use. This Apple butter is not at all like Apple sauce; it has no granulated appearance, but is smooth and solid like cheese and dark red in color. Apple butter is stirred by a pole having upon one end a perforated blade or paddle set at right angles. Sometimes a bar was laid from rim to rim of the caldron, and worked by a crank that turned a similar paddle. A collection of ancient utensils used in making Apple butter is shown on page 211; these are from the collections of the Bucks County Historical Society. Opposite page 214 is shown an ancient open-air fireplace and an old couple making Apple butter just as they have done for over half a century.

In New England what the "hired man" on the farm called "biled cider Apple sass," took the place of Apple butter. Preferably this was made in the "summer kitchen," where three kettles, usually of graduated sizes, could be set over the fire; the three kettles could be hung from a crane, or trammels. All were filled with cider, and as the liquid boiled away in the largest kettle it was filled from the second and that from the third. The fresh cider was always poured into the third kettle, thus the large kettle was never checked in its boiling. This continued till the cider was as thick as molasses. Apples (preferably Pound Sweets or Pumpkin Sweets) had been chosen with care, pared, cored, and quartered, and heated in a small kettle. These were slowly added to the thickened cider, in small quantities, in order not to check the boiling. The rule was to cook them till so softened that a rye straw could be run into them, and yet they must retain their shape. This was truly a critical time; the slightest scorched flavor would ruin the whole kettleful. A great wooden, long-handled, shovel-like ladle was used to stir the sauce fiercely until it was finished in triumph. Often a barrel of this was made by our grandmothers, and frozen solid for winter use. The farmer and "hired men" ate it clear as a relish with meats; and it was suited to appetites and digestions which had been formed by a diet of salted meats, fried breads, many pickles, and the drinking of hot cider sprinkled with pepper.

Emerson well named the Apple the social fruit of New England. It ever has been and is still the grateful promoter and unfailing aid to informal social intercourse in the country-side; but the Apple tree is something far nobler even than being the sign of cheerful and cordial acquaintance; it is the beautiful rural emblem of industrious and temperate home life. Hence, let us wassail with a will: —

> "Here's to thee, old Apple tree!
> Whence thou mayst bud, and whence thou mayst blow,
> And whence thou mayst bear Apples enow!"

Making Apple Butter.

CHAPTER IX

GARDENS OF THE POETS

"The chief use of flowers is to illustrate quotations from the poets."

LL English poets have ever been ready to sing English flowers until jesters have laughed, and to sing garden flowers as well as wild flowers. Few have really described a garden, though the orderly distribution of flowers might be held to be akin to the restraint of rhyme and rhythm in poetry.

It has been the affectionate tribute and happy diversion of those who love both poetry and flowers to note the flowers beloved of various poets, and gather them together, either in a book or a garden. The pages of Milton cannot be forced, even by his most ardent admirers, to indicate any intimate knowledge of flowers. He certainly makes some very elegant classical allusions to flowers and fruits, and some amusingly vague ones as well. "The Flowers of Spenscr," and "A Posy from Chaucer," are the titles of most readable chapters in *A Garden of Simples*, but the allusions and quotations from both authors are pleasing and interesting, rather than informing as to the real variety and description of the flowers of their day. Nearly all the older English poets, though writing glibly of woods and vales, of shepherds and swains, of buds and blossoms, scarcely allude to a flower in a natural way. Herrick was truly a flower lover, and, as the critic said, "many flowers grow to illustrate

Shakespeare Border at Hillside.

quotations from his works." The flowers named of Shakespeare have been written about in varied books, *Shakespeare's Garden*, *Shakespeare's Bouquet*, *Flowers from Stratford-on-Avon*, etc. These are easily led in fulness of detail, exactness of information, and delightful literary quality by that truly perfect book, beloved of all garden lovers, *The Plant Lore and Garden Craft of Shakespeare*, by Canon Ellacombe. Of it I never weary, and for it I am ever grateful.

Shakespeare Gardens, or Shakespeare Borders, too, are laid out and set with every tree, shrub, and flower named in Shakespeare, and these are over two hundred in number. A distinguishing mark of the Shakespeare Border of Lady Warwick is the peculiar label set alongside each plant. This label is of pottery, greenish-brown in tint, shaped like a butterfly, bearing on its wings a quotation of a few words and the play reference relating to each special plant. Of course these words have been fired in and are thus permanent. Pretty as they are in themselves they must be disfiguring to the borders — as all labels are in a garden.

In the garden at Hillside, near Albany, New York, grows a green and flourishing Shakespeare Border, gathered ten years ago by the mistress of the garden. I use the terms green and flourishing with exactness in this connection, for a great impression made by this border is of its thriving health, and also of the predominance of green leafage of every variety, shape, manner of growth, and oddness of tint. In this latter respect it is infinitely more beautiful than the ordinary border, varying from silvery glaucous green through greens of yellow or brownish shade to the blue-black greens of some herbs; and among these green leaves are many of sweet or pungent scent, and of medicinal qualities, such as are seldom grown to-day save in some such choice and chosen spot. There is less bloom in this Shakespeare Border than in our modern flower beds, and the flowers are not so large or brilliant as our modern favorites; but, quiet as they are, they are said to excel the blossoms of the same plants of Shakespeare's own day, which we learn from the old herbalists were smaller and less varied in color and of simpler tints than those of their descendants. At the first glance this Shakespeare Border shines chiefly in the light of the imagination, as stirred by the poet's noble words; but do not dwell on this border as a whole, as something only to be looked at; read the pages of this garden, dwell on each leafy sentence, and you are entranced with its beautiful significance. It was not gathered with so much thought, and each plant and seed set out and watched and reared like a delicate child, to become a show place; it appeals for a more intimate regard; and we find that its detail makes its charm.

Such a garden as this appeals warmly to anyone who is sensitive to the imaginative element of flower beauty. Many garden makers forget that a flower bed is a group of living beings — perhaps of sentient beings — as well as a mass of beautiful color. Modern gardens tend far too much toward the display of the united effect of growing plants, to a striving for universal brilliancy, rather than attention to and love for separate flowers. There was refreshment of spirit as well as of the senses in the old-time garden of flowers, such as these planted in this Shakespeare Border, and it stirred the heart of the poet as could no modern flower gardens.

The scattering inflorescence and the tiny size of the blossoms give to this Shakespeare Border an

Long Border at Hillside.

unusual aspect of demureness and delicacy, and the plants seem to cling with affection and trust to the path of their human protector; they look simple and confiding, and seem close both to nature and to man. This homelike and modest quality is shown, I think, even in the presentation in black and white given on page 216 and opposite page, 218, though it shows still more in the garden when the wide range of tint of foliage is added.

A most appropriate companion of the old flowers in this Shakespeare Border is the sun-dial, which is an exact copy of the one at Abbotsford, Scotland. It bears the motto 'ΕΡΧΕΤΑΙ ΓΑΡ ΝΥΞ meaning, "For the night cometh." It was chosen by Sir Walter Scott, for his sun-dial, as a solemn monitor to himself of the hour "when no man can work." It was copied from a motto on the dial-plate of the watch of the great Dr. Samuel Johnson; and it is curious that in both cases the word ΓΑΡ should be introduced, for it is not in the clause in the New Testament from which the motto was taken. It is a beautiful motto and one of singular appropriateness for a sun-dial. The pedestal of this sun-dial is of simple lines, but it is dignified and pleasing, aside from the great interest of association which surrounds it.

I had a happy sense, when walking through this garden, that, besides my congenial living companionship, I had the company of some noble Elizabethan ghosts; and I know that if Shakespeare and Jonson and Herrick were to come to Hillside, they would find the garden so familiar to them; they would

The Beauty of Winter Lilacs.

greet the plants like old friends, they would note how fine grew the Rosemary this year, how sweet were the Lady's-smocks, how fair the Gillyflowers. And Gerarde and Parkinson would ponder, too, over all the herbs and simples of their own Physick Gardens, and compare notes. Above all I seemed to see, walking soberly by my side, breathing in with delight the varied scents of leaf and blossom, that lover and writer of flowers and gardens, Lord

Bacon — and not in the disguise of Shakespeare either. For no stronger proofs can be found of the existence of two individualities than are in the works of each of these men, in their sentences and pages which relate to gardens and flowers.

This fair garden and Shakespeare Border are loveliest in the cool of the day, in the dawn or at early eve; and those who muse may then remember another Presence in a garden in the cool of the day. And then I recall that gem of English poesy which always makes me pitiful of its author; that he could write this, and yet, in his hundreds of pages of English verse, make not another memorable line: —

"A Garden is a lovesome thing, God wot;
Rose plot,
Fringed pool,
Ferned grot,
The veriest school of Peace;
And yet the fool
Contends that God is not in gardens.
Not in gardens! When the eve is cool!
Nay, but I have a sign.
'Tis very sure God walks in mine."

Shakespeare Borders grow very readily and freely in England, save in the case of the few tropical flowers and trees named in the pages of the great dramatist; but this Shakespeare Border at Hillside needs much cherishing. The plants of Heather and Broom and Gorse have to be specially coddled by transplanting under cold frames during the long winter months in frozen Albany; and thus they find vast contrast to their free, unsheltered life in Great Britain.

Persistent efforts have been made to acclimate both Heather and Gorse in America. We have seen how Broom came uninvited and spread unasked on the Massachusetts coast; but Gorse and Heather have proved shy creatures. On the beautiful island of Naushon the carefully planted Gorse may be found spread in widely scattered spots and also on the near-by mainland, but it cannot be said to have

Garden of Mrs. Frank Robinson, Wakefield, Rhode Island.

thrived markedly. The Scotch Heather, too, has been frequently planted, and watched and pushed, but it is slow to become acclimated. It is not because the winters are too cold, for it is found in considerable amount in bitter Newfoundland; perhaps it prefers to live under a crown.

Modern authors have seldom given their names to gardens, not even Tennyson with his intimate and extended knowledge of garden flowers. A

Mary Howitt Garden was planned, full of homely old blooms, such as she loves to name in her verse; but it would have slight significance save to its maker, since no one cares to read Mary Howitt nowadays. In that charming book, *Sylvana's Letters to an Unknown Friend* (which I know were written to me), the author, E. V. B., says, "The very ideal of a garden, and the only one I know, is found in Shelley's *Sensitive Plant*." With quick championing of a beloved poet, I at once thought of the radiant garden of flowers in Keats's heart and poems. Then I reread the *Sensitive Plant* in a spirit of utmost fairness and critical friendliness, and I am willing to yield the Shelley Garden to Sylvana, while I keep, for my own delight, my Keats garden of sunshine, color, and warmth.

That Keats had a profound knowledge and love of flowers is shown in his letters as well as his poems. Only a few months before his death, when stricken with and fighting a fatal disease, he wrote: —

"How astonishingly does the chance of leaving the world impress a sense of its natural beauties upon me! Like poor Falstaff, though I do not babble, I think of green fields. I muse with greatest affection on every flower I have known from my infancy — their shapes and colors are as new to me as if I had just created them with a superhuman fancy. It is because they are connected with the most thoughtless and the happiest moments of my life."

Near the close of his *Endymion* he wrote: —

"Nor much it grieves
To die, when summer dies on the cold sward.
Why, I have been a butterfly, a lord
Of flowers, garlands, love-knots, silly posies,
Groves, meadows, melodies, and arbor roses;
My kingdom's at its death, and just it is
That I should die with it."

In the summer of 1816, under the influence of a happy day at Hampstead, he wrote that lovely poem, "I stood tiptoe upon a little hill." After a description of the general scene, a special corner of beauty is thus told: —

"A bush of May flowers with the bees about them —
Ah, sure no bashful nook could be without them —
And let a lush Laburnum oversweep them,
And let long grass grow round the roots to keep them
Moist, cool, and green; and shade the Violets
That they may bind the moss in leafy nets.
A Filbert hedge with Wild-brier over trim'd,
And clumps of Woodbine taking the soft wind,
Upon their summer thrones. . . ."

Then come these wonderful lines, which belittle all other descriptions of Sweet Peas: —

"Here are Sweet Peas, on tiptoe for a flight,
With wings of gentle flush o'er delicate white,
And taper fingers catching at all things
To bind them all about with tiny wings."

Keats states in his letters that his love of flowers was wholly for those of the "common garden sort," not for flowers of the greenhouse or difficult cultivation, nor do I find in his lines any evidence

The Parson's Walk.

of extended familiarity with English wild flowers. He certainly does not know the flowers of woods and fields as does Matthew Arnold.

The Autocrat of the Breakfast Table says: "Did you ever hear a poet who did not talk flowers? Don't you think a poem which for the sake of being original should leave them out, would be like those verses where the letter *a* or *e*, or some other, is omitted? No; they will bloom over and over again in poems as in the summer fields, to the end of time, always old and always new." The Autocrat himself knew well a poet who never talked flowers in his poems, a poet beloved of all other poets, — Arthur Hugh Clough, — though he loved and knew all flowers. From Matthew Arnold's beautiful tribute to him, are a few of his wonderful flower lines, cut out from their fellows: —

> "Through the thick Corn the scarlet Poppies peep,
> And round green roots and yellowing stalks I see
> Pale blue Convolvulus in tendrils creep,
> And air-swept Lindens yield
> Their scent, and rustle down their perfumed showers
> Of bloom. . . ,
>
> * * * * *
>
> "Soon will the high midsummer pomps come on,
> Soon will the Musk Carnations break and swell.
> Soon shall we have gold-dusted Snapdragon,
> Sweet-william with his homely cottage smell,
> And Stocks in fragrant blow."

Oh, what a master hand! Where in all English verse are fairer flower hues? And where is a more beautiful description of a midsummer evening, than Arnold's exquisite lines beginning: —

> "The evening comes; the fields are still;
> The tinkle of the thirsty rill."

Dr. Holmes was also a master in the description of garden flowers. I should know, had I never been told save from his verses, just the kind of a Cambridge garden he was reared in, and what flowers grew in it. Lowell, too, gives ample evidence of a New England childhood in a garden.

The gardens of Shenstone's *Schoolmistress* and of Thomson's poems come to our minds without great warmth of welcome from us; while Clare's lines are full of charm: —

> "And where the Marjoram once, and Sage and Rue,
> And Balm, and Mint, with curl'd leaf Parsley grew,
> And double Marigolds, and silver Thyme,
> And Pumpkins 'neath the window climb.
> And where I often, when a child, for hours
> Tried through the pales to get the tempting flowers,
> As Lady's Laces, everlasting Peas,
> True-love-lies-bleeding, with the Hearts-at-ease
> And Goldenrods, and Tansy running high,
> That o'er the pale tops smiled on passers by."

A curious old seventeenth-century poet was the Jesuit, René Rapin. The copy of his poem entitled *Gardens* which I have seen, is the one in my daughter's collection of garden books; it was "English'd by the Ingenious Mr. Gardiner," and published in 1728. Hallam in his *Introduction to the Literature of Europe* gives a capital estimate of this long poem of over three thousand lines. I find them pretty dull reading, with much monotony of adjectives, and very affected notions for plant names. I fancy he manufactured all his tedious plant traditions himself.

A pleasing little book entitled *Dante's Garden* has collected evidence, from his writings, of Dante's love of green, growing things. The title is rather strained, since he rarely names individual flowers, and only refers vaguely to their emblematic significance. I would have entitled the book *Dante's Forest*, since he chiefly refers to trees; and the Italian gardens of his days were of trees rather than flowers. There are passages in his writings which have led some of his worshippers to believe that his childhood was passed in a garden; but these references are very indeterminate. The picture of a deserted garden, with its sad sentiment has charmed the fancy of many a poet. Hood, a true flower-lover, wrote this jingle in his *Haunted House*: —

Garden of Mary Washington.

"The Marigold amidst the nettles blew,
 The Gourd embrac'd the Rose bush in its ramble.
The Thistle and the Stock together grew,
 The Hollyhock and Bramble.

"The Bearbine with the Lilac interlaced,
The sturdy Burdock choked its tender neighbor,
The spicy Pink. All tokens were effaced
Of human care and labor."

These lines are a great contrast to the dignified versification of The Old Garden, by Margaret Deland, a garden around which a great city has grown.

"Around it is the street, a restless arm
That clasps the country to the city's heart."

No one could read this poem without knowing that the author is a true garden lover, and knowing as well that she spent her childhood in a garden.

Another American poet, Edith Thomas, writes exquisitely of old gardens and garden flowers.

"The pensile Lilacs still their favors throw.
The Star of Lilies, plenteous long ago,
Waits on the summer dusk, and faileth not.
The legions of the grass in vain would blot
The spicy Box that marks the garden row.
Let but the ground some human tendance know,
It long remaineth an engentled spot."

Let me for a moment, through the suggestion of her last two lines, write of the impress left on nature through flower planting. "The garden long remaineth an engentled spot." You cannot for years stamp out the mark of a garden; intentional destruction may obliterate the garden borders, but neglect never. The delicate flowers die, but some sturdy things spring up happily and seem gifted with everlasting life. Fifteen years ago a friend bought an old country seat on Long Island; near the site of the new house, an old garden was ploughed deep and levelled to a lawn. Every year since then the patient gardeners pull up, on this lawn, in considerable numbers, Mallows, Campanulas, Star of Bethlehem, Bouncing-bets and innumerable Asparagus shoots, and occasionally the seedlings of other flowers which have bided their time in the dark earth. Traces of the residence of Sir Walter Raleigh in Ireland may still be seen in the growth of richly perfumed wall-flowers which he brought from the Azores. The Affane Cherry is found where he planted it, and some of his Cedars are living. The summer-house of Yew trees sheltered him when he smoked in the garden, and in this garden he planted Tobacco. Near by is the famous spot where he planted what were then called Virginian Potatoes. By that planting they acquired the name of Irish Potatoes.

I have spoken of the Prince Nurseries in Flushing; the old nurserymen left a more lasting mark than their Nurseries, in the rare trees and plants now found on the roads, and in the fields and gardens for many miles around Flushing. With the Parsons family, who have been, since 1838, distributors of unusual plants, especially the splendid garden treasures from China and Japan, they have made Flushing a delightful nature-study.

In the humblest dooryard, and by the wayside in outlying parts of the town, may be seen rare and beautiful old trees: a giant purple Beech is in a laborer's yard; fine Cedars, Salisburias, red-flowered Horse-chestnuts, Japanese flowering Quinces and

Cherries, and even rare Japanese Maples are to be found; a few survivors of the Chinese Mulberry have a romantic interest as mementoes of a giant bubble of ruin. The largest Scotch Laburnum I ever saw, glorious in golden bloom, is behind an unkempt house. On the Parsons estate is a weeping Beech of unusual size. Its branches trail on

Box and Phlox.

the ground in a vast circumference of 222 feet, forming a great natural arbor. The beautiful vernal light in this tree bower may be described in Andrew Marvell's words:—

"Annihilating all that's made
To a green thought in a green shade."

The photograph of it, shown opposite page 232, gives some scant idea of its leafy walls; it has been for years the fit trysting-place of lovers, as is shown by the initials carved on the great trunk. Great

Judas trees, sadly broken yet bravely blooming; decayed hedges of several kinds of Lilacs, Syringas, Snowballs, and Yuccas of princely size and bearing still linger. Everywhere are remnants of Box hedges. One unkempt dooryard of an old Dutch farm-house was glorified with a broad double row of yellow Lily at least sixty feet in length. Everywhere is Wistaria, on porches, fences, houses, and trees; the abundant Dogwood trees are often overgrown with Wistaria. The most exquisite sight of the floral year was the largest Dogwood tree I have ever seen, radiant with starry white bloom, and hung to the tip of every white-flowered branch with the drooping amethystine racemes of Wistaria of equal luxuriance. Golden-yellow Laburnum blooms were in one case mingled with both purple and white Wistaria. These yellow, purple, and white blooms of similar shape were a curious sight, as if a single plant had been grafted. As I rode past so many glimpses of loveliness mingled with so much present squalor, I could but think of words of the old hymn: —

> "Where every prospect pleases
> And only man is vile."

Could the hedges, trees, and vines which came from the Prince and Parsons Nurseries have been cared for, northeastern Long Island, which is part of the city of Greater New York, would still be what it was named by the early explorers, "The Pearl of New Netherland."

Within the Weeping Beech.

CHAPTER X

THE CHARM OF COLOR

> "How strange are the freaks of memory,
> The lessons of life we forget.
> While a trifle, a trick of color,
> In the wonderful web is set."
>
> —JAMES RUSSELL LOWELL.

HE quality of charm in color is most subtle; it is like the human attribute known as fascination, "whereof," says old Cotton Mather, "men have more Experience than Comprehension." Certainly some alliance of color with a form suited or wonted to it is necessary to produce a gratification of the senses. Thus in the leaves of plants every shade of green is pleasing; then why is there no charm in a green flower? The green of Mignonette bloom would scarcely be deemed beautiful were it not for our association of it with the delicious fragrance. White is the absence of color. In flowers a pure chalk-white, and a snow-white (which is bluish) is often found; but more frequently the white flower blushes a little, or is warmed with yellow, or has green veins.

Where green runs into the petals of a white flower, its beauty hangs by a slender thread. If the green lines have any significance, as have the faint green checkerings of the Fritillary, which I have described elsewhere in this book, they add to its interest; but ordinarily they make the petals seem undeveloped. The Snowdrop bears the mark of one of the few tints of green which we like in white flowers; its "heart-shaped seal of green,"

Spring Snowflake.

sung by Rossetti, has been noted by many other poets. Tennyson wrote:—

> "Pure as lines of green that streak the white
> Of the first Snowdrop's inner leaves."

A cousin of the Snowdrop, is the "Spring Snowflake" or Leucojum, called also by New England country folk "High Snowdrop." It bears at the end of each snowy petal a tiny exact spot of green;

and I think it must have been the flower sung by Leigh Hunt: —

> "The nice-leaved lesser Lilies,
> Shading like detected light
> Their little green-tipt lamps of white."

The illustration on page 234 shows the graceful growth of the flower and its exquisitely precise little green-dotted petals, but it has not caught its luminous whiteness, which seems almost of phosphorescent brightness in each little flower.

The Star of Bethlehem is a plant in which the white and green of the leaf is curiously repeated in the flower. Gardeners seldom admit this flower now to their gardens, it so quickly crowds out everything else; it has become on Long Island nothing but a weed. The high-growing Star of Bethlehem is a pretty thing. A bed of it in my sister's garden is shown on page 237.

It is curious that when all agree that green flowers have no beauty and scant charm, that a green flower should have been one of the best-loved flowers of my home garden. But this love does not come from any thought of the color or beauty of the flower, but from association. It was my mother's favorite, hence it is mine. It was her favorite because she loved its clear, pure, spicy fragrance. This ever present and ever welcome scent which pervades the entire garden if leaf or flower of the loved Ambrosia be crushed, is curious and characteristic, a true "ambrosiack odor," to use Ben Jonson's words.

A vivid description of Ambrosia is that of Gerarde in his delightful *Herball.*

"Oke of Jerusalem, or Botrys, hath sundry small stems a foote and a halfe high dividing themselves into many small branches. The leafe very much resembling the leafe of an Oke, which hath caused our English women to call it Oke of Jerusalem. The upper side of the leafe is a deepe greene and somewhat rough and hairy, but underneath it is of a darke reddish or purple colour. The seedie floures grow clustering about the branches like the yong clusters or blowings of the Vine. The roote is small and thriddy. The whole herbe is of a pleasant smell and savour, and the whole plant dieth when the seed is ripe. Oke of Jerusalem is of divers called Ambrosia."

Ambrosia has been loved for many centuries by Englishwomen; it is in the first English list of names of plants, which was made in 1548 by one Dr. Turner; and in this list it is called "Ambrose." He says of it:—

"Botrys is called in englishe, Oke of Hierusalem, in duche, trauben kraute, in french pijmen. It groweth in gardines muche in England."

Ambrosia has now died out "in gardines muche in England." I have had many letters from English flower lovers telling me they know it not; and I have had the pleasure of sending the seeds to several old English and Scotch gardens, where I hope it will once more grow and flourish, for I am sure it must feel at home.

The seeds of this beloved Ambrosia, which filled my mother's garden in every spot in which it could spring, and which overflowed with cheerful welcome into the gardens of our neighbors, was given her from the garden of a great-aunt in Walpole, New Hampshire. This Walpole garden was

Star of Bethlehem.

a famous gathering of old-time favorites, and it had the delightful companionship of a wild garden. On a series of terraces with shelving banks, which reached down to a stream, the boys of the family planted, seventy years ago, a myriad of wild flowers, shrubs, and trees, from the neighboring woods. By the side of the garden great Elm trees sheltered scores of beautiful gray squirrels; and behind the house and garden an orchard led to the wheat fields, which stretched down to the broad Connecticut River. All flowers thrived there, both in the Box-bordered beds and in the wild garden, perhaps because the morning mists from the river helped out the heavy buckets of water from the well during the hot summer

"The Pearl."

weeks. Even in winter the wild garden was beautiful from the brilliant Bittersweet which hung from every tree.

Here Ambrosia was plentiful, but is plentiful no longer; and Walpole garden lovers seek seeds of it from the Worcester garden. I think it dies out generally when all the weeding and garden care is done by gardeners; they assume that the little plants of such modest bearing are weeds, and pull them up, with many other precious seedlings of the old garden, in their desire to have ample expanse of naked dirt. One of the charms which was permitted to the old garden was its fulness. Nature there certainly abhorred a vacant space. The garden soil was full of resources; it had a seed for every square inch; it seemed to have a reserve store ready to crowd into any space offered by the removal or dying down of a plant at any time.

Let me tell of a curious thing I found in an old book, anent our subject — green flowers. It shows that we must not accuse our modern sensation lovers, either in botany or any other science, of being the only ones to add artifice to nature. The green Carnation has been chosen to typify the decadence and monstrosity of the end of the nineteenth century; but nearly two hundred years ago a London fruit and flower grower, named Richard Bradley, wrote a treatise upon field husbandry and garden culture, and in it he tells of a green Carnation which "a certayn fryar" produced by grafting a Carnation upon a Fennel stalk. The flowers were green for several years, then nature overcame decadent art.

There be those who are so enamoured of the color green and of foliage, that they care little for flowers of varied tint; even in a garden, like the old poet Marvell, they deem, —

"No white nor red was ever seen
So amorous as this lovely green."

Such folk could scarce find content in an American garden; for our American gardeners must confess, with Shakespeare's clown: "I am no great Nebuchadnezzar, sir, I have not much skill in grass." Our lawns are not old enough.

A charming greenery of old English gardens was the bowling-green. We once had them in our colonies, as the name of a street in our greatest city now proves; and I deem them a garden fashion well-to-be-revived.

The laws of color preference differ with the size of expanses. Our broad fields often have pleasing expanses of leafage other than green, and flowers that are as all-pervading as foliage. Many flowers of the field have their day, when each seems to be queen, a short day, but its rights none dispute. Snow of Daisies, yellow of Dandelions, gold of Buttercups, purple pinkness of Clover, Innocence, Blue-eyed Grass, Milkweed, none reign more absolutely in every inch of the fields than that poverty stricken creature, the Sorrel. William Morris warns us that "flowers in masses are mighty strong color," and must be used with much caution in a garden. But there need be no fear of massed color in a field, as being ever gaudy or cloying. An approach to the beauty and satisfaction of nature's plentiful field may be artificially obtained as an adjunct to the garden in a flower-close sown or set with a solid expanse of bloom of some native or widely adopted plant. I have seen a flower-close of Daisies, another of Buttercups, one of Larkspur, one of Coreopsis. A new field tint, and a splendid one, has been given to us within a few years, by the introduction of the vivid red of Italian clover. It is eagerly welcomed to our fields, so scant of scarlet. This clover was brought to America in the years 1824 *et seq.*, and is described in contemporary publications in alluring sentences. I have noted the introduction of several vegetables, grains, fruits, berries, shrubs, and flowers in those years, and attribute this to the influence of the visit of Lafayette in 1824. Adored by all, his lightest word was heeded; and he was a devoted agriculturist and horticulturist, ever exchanging ideas, seeds, and plants with his American fellow-patriots and fellow-farmers. I doubt if Italian clover then became widely known; but our modern farmers now think well of it, and the flower lover revels in it.

The exigencies of rhyme and rhythm force us to endure some very curious notions of color in the poets. I think no saying of poet ever gave greater check to her lovers than these lines of Emily Dickinson: —

> "Nature rarer uses yellow
> Than another hue;
> Saves she all of that for sunsets,
> Prodigal of blue.
> Spending scarlet like a woman,
> Yellow she affords
> Only scantly and selectly,
> Like a lover's words."

I read them first with a sense of misapprehension that I had not seen aright; but there the words stood out, "Nature rarer uses yellow than another hue." The writer was such a jester, such a tricky

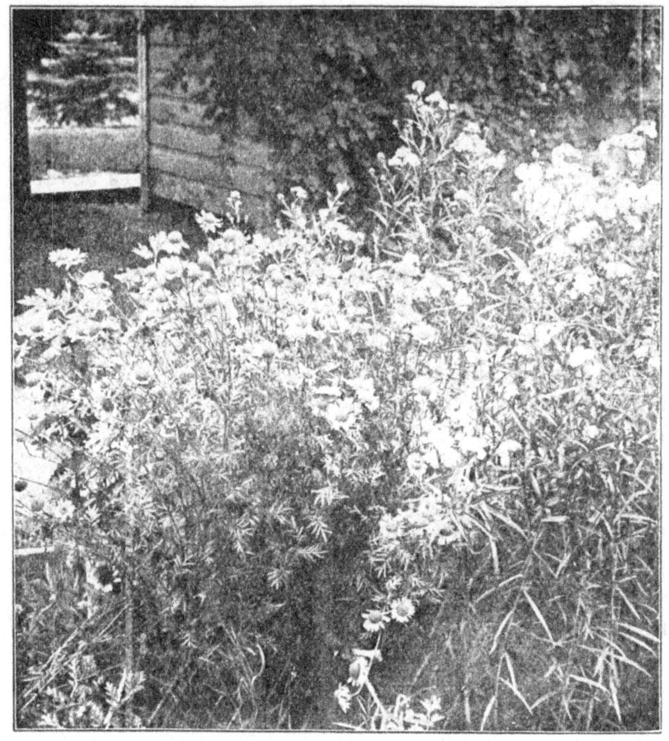

Pyrethrum.

elf that I fancy she wrote them in pure "contrariness," just to see what folks would say, how they would dispute over her words. For I never can doubt that, with all her recluse life, she knew intuitively that some time her lines would be read by folks who would love them.

The scarcity of red wild flowers is either a cause

or an effect; at any rate it is said to be connected with the small number of humming-birds, who play an important part in the fertilization of many of the red flowers. There are no humming-birds in Europe; and the Aquilegia, red and yellow here, is blue there, and is then fertilized by the assistance of the bumblebee. Without humming-birds the English successfully accomplish one glorious sweep of red in the Poppies of the field; Parkinson called them "a beautiful and gallant red" — a very happy phrase. Ruskin, that master of color and of its description, and above all master of the description of Poppies, says: —

> "The Poppy is the most transparent and delicate of all the blossoms of the field. The rest, nearly all of them, depend on the texture of their surface for color. But the Poppy is painted glass; it never glows so brightly as when the sun shines through it. Whenever it is seen, against the light or with the light, it is a flame, and warms the wind like a blown ruby."

There is one quality of the Oriental Poppies which is very palpable to me. They have often been called insolent — Browning writes of the "Poppy's red affrontery"; to me the Poppy has an angry look. It is wonderfully haughty too, and its seed-pod seems like an emblem of its rank. This great green seed-pod stands one inch high in the centre of the silken scarlet robe, and has an antique crown of purple bands with filling of lilac, just like the crown in some ancient kingly portraits, when the bands of gold and gems radiating from a great jewel in the centre are filled with crimson or purple velvet. Around this splendid crowned seed-vessel are rows of stamens and purple anthers of richest hue.

We must not let any scarlet flower be dropped from the garden, certainly not the Geranium, which just at present does not shine so bravely as a few years ago. The general revulsion of feeling against "bedding out" has extended to the poor plants thus misused, which is unjust. I find I have spoken somewhat despitefully of the Coleus, Lobelia, and Calceolaria, so I hasten to say that I do not include the Geranium with them. I love its clean color, in leaf and blossom; its clean fragrance; its clean beauty, its healthy growth; it is a plant I like to have near me.

It has been the custom of late to sneer at crimson in the garden, especially if its vivid color gets a dash of purple and becomes what Miss Jekyll calls "malignant magenta." It is really more vulgar than malignant, and has come to be in textile products a stamp and symbol of vulgarity, through the forceful brilliancy of our modern aniline dyes. But this purple crimson, this amarant, this magenta, especially in the lighter shades, is a favorite color in nature. The garden is never weary of wearing it. See how it stands out in midsummer! It is rank in Ragged Robin, tall Phlox, and Petunias; you find it in the bed of Drummond Phlox, among the Zinnias; the Portulacas, Balsams, and China Asters prolong it. Earlier in the summer the Rhododendrons fill the garden with color that on some of the bushes is termed sultana and crimson, but it is in fact plain magenta. One of the good points of the Peony is that you never saw a magenta one.

This color shows that time as well as place affects our color notions, for magenta is believed to be the honored royal purple of the ancients. Fifty years ago no one complained of magenta. It was deemed a cheerful color, and was set out boldly and complacently by the side of pink or scarlet, or wall flower colors. Now I dislike it so that really the printed word, seen often as I glance back through this page, makes the black and white look cheap. If I could turn all magenta flowers pink or purple, I should never think further about garden harmony, all other colors would adjust themselves.

It has been the fortune of some communities to be the home of men in nature like Thoreau of Concord and Gilbert White of Selborne, men who live solely in love of out-door things, birds, flowers, rocks, and trees. To all these nature lovers is not given the power of writing down readily what they see and know, usually the gift of composition is denied them; but often they are just as close and accurate observers as the men whose names are known to the world by their writings. Sometimes these naturalists boldly turn to nature, their loved mother, and earn their living in the woods and fields. Sometimes they have a touch of the hermit in them, they prefer nature to man; others are genial, kindly men, albeit possessed of a certain reserve. I deem the community blest that has such a citizen, for his influence in promoting a love and study of nature is ever great. I have known one such ardent naturalist, Arba Peirce, ever since my childhood. He lives the greater part of

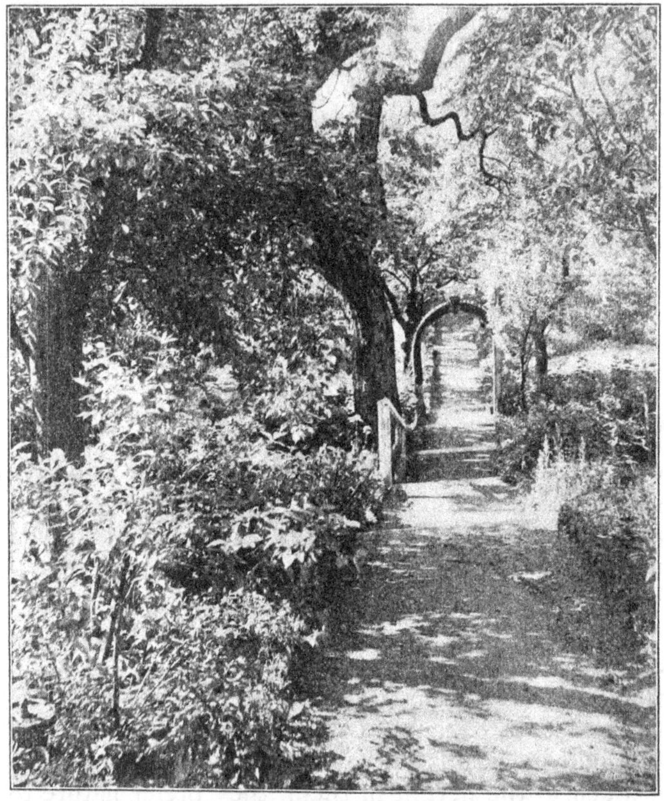

Terraced Garden of Misses Nichols, Salem, Massachusetts.

his waking hours in the woods and fields, and these waking hours are from sunrise. From the earliest bloom of spring to the gay berry of autumn, he knows all beautiful things that grow, and where they grow, for hundreds of miles around his home.

I speak of him in this connection because he has acquired through his woodland life a wonderful power of distinguishing flowers at great distance with absolute accuracy. Especially do his eyes have the power of detecting those rose-lilac tints which are characteristic of our rarest, our most delicate wild flowers, and which I always designate to myself as Arethusa color. He brought me this June a royal gift — a great bunch of wild fringed Orchids, another of Calopogon, and one of Arethusa. What a color study these three made! At the time their lilac-rose tints seemed to me far lovelier than any pure rose colors. In those wild princesses were found every tone of that lilac-rose from the faint blush like the clouds of a warm sunset, to a glow on the lip of the Arethusa, like the crimson glow of Mullein Pink.

My friend of the meadow and wildwood had gathered that morning a glorious harvest, over two thousand stems of Pogonia, from his own hidden spot, which he has known for forty years and from whence no other hand ever gathers. For a little handful of these flower heads he easily obtains a dollar. He has acquired gradually a regular round of customers, for whom he gathers a successive harvest of wild flowers from Pussy Willows and Hepatica to winter berries. It is not easily earned money to stand in heavy rubber boots in marsh mud and water reaching nearly to the waist, but after all it is happy work. Jeered at in his early life by fools for his wood-roving tastes, he has now the pleasure and honor of supplying wild flowers to our public schools, and being the authority to whom scholars and teachers refer in vexed questions of botany.

I think the various tints allied to purple are the most difficult to define and describe of any in the garden. To begin with, all these pinky-purple, these arethusa tints are nameless; perhaps orchid color is as good a name as any. Many deem purple and violet precisely the same. Lavender has much gray in its tint. Miss Jekyll deems mauve and lilac the same; to me lilac is much pinker, much more delicate. Is heliotrope a pale bluish purple? Some call it a blue faintly tinged with red. Then there are the orchid tints, which have more pink than blue. It is a curious fact that, with all these allied tints which come from the union of blue with red, the color name comes from a flower name. Violet, lavender, lilac, heliotrope, orchid, are examples; each is an exact tint. Rose and pink are color names from flowers, and flowers of much variety of colors, but the tint name is unvarying.

Edward de Goncourt, of all writers on flowers and gardens, seems to have been most frankly pleased with the artificial side of the gardener's art. He viewed the garden with the eye of a colorist, setting a palette of varied greens from the deep tones of the evergreens, the Junipers and Cryptomerias through the variegated Hollies, Privets and Spindle trees;

and he said that an "elegantly branched coquettishly variegated bush" seemed to him like a piece of bric-a-brac which should be hunted out and praised like some curio hidden on the shelf of a collector.

A lack of color perception seems to have been prevalent of ancient days, as it is now in some Oriental countries. The Bible offers evidence of this, and it has also been observed that the fragrance of flowers is nowhere noted until we reach the Song of Solomon. It is believed that in earliest time archaic men had no sense of color; that they knew only light and darkness. Mr. Gladstone wrote a most interesting paper on the lack of color sense in Homer, whose perception of brilliant light was good, especially in the glowing reflections of metals, but who never names blue or green even in speaking of the sky, or trees, while his reds and purples are hopelessly mixed. Some German scientists have maintained that as recently as Homer's day, our ancestors were (to use Sir John Lubbock's word) blue-blind, which fills me, as it must all blue lovers, with profound pity.

The influence of color has ever been felt by other senses than that of sight. In the *Cotton Manuscripts*, written six hundred years ago, the relations and effects of color on music and coat-armor were laboriously explained: and many later writers have striven to show the effect of color on the health, imagination, or fortune. I see no reason for sneering at these notions of sense-relation; I am grateful for borrowed terms of definition for these beautiful things which are so hard to define. When an artist says to me, "There is a color that sings," I know what he means; as I do when my friend says of the funeral music in *Tristan* that "it always hurts her eyes." Musicians compose symphonies in color, and artists paint pictures in symphonies. Musicians and authors

Arbor in a Salem Garden.

acknowledge the domination of color and color terms; a glance at a modern book catalogue will prove it. Stephen Crane and other modern extremists depend upon color to define and describe sounds, smells, tastes, feelings, ideas, vices, virtues, traits, as well as sights. Sulphur-yellow is deemed an inspiring color, and light green a clean color; every one knows the influence of bright red upon many animals and birds; it is said all barnyard fowl are affected by it. If any one can see a sunny bed of blue Larkspur in full bloom without being moved thereby, he must be color blind and sound deaf as well, for that indeed is a sight full of music and noble inspiration, a realization of Keats' beautiful thought: —

> "Delicious symphonies, like airy flowers
> Budded, and swell'd, and full-blown, shed full showers
> Of light, soft unseen leaves of sound divine."

CHAPTER XI

THE BLUE FLOWER BORDER

"Blue thou art, intensely blue!
Flower! whence came thy dazzling hue?
When I opened first mine eye,
Upward glancing to the sky,
Straightway from the firmament
Was the sapphire brilliance sent."

—James Montgomery.

UESTIONS of color relations in a garden are most opinion-making and controversy-provoking. Shall we plant by chance, or by a flower-loving instinct for sheltered and suited locations, as was done in all old-time gardens, and with most happy and most unaffected results? or shall we plant severely by colors — all yellow flowers in a border together? all red flowers side by side? all pink flowers near each other? This might be satisfactory in small gardens, but I am uncertain whether any profound gratification or full flower succession would come from such rigid planting in long flower borders.

William Morris warns us that flowers in masses are "mighty strong color," and must be used with caution. A still greater cause for hesitation would be the ugly jarring of juxtaposing tints of the same color. Yellows do little injury to each other; but I cannot believe that a mixed border of red flowers would ever be satisfactory or scarcely endurable; and few persons would care for beds of all white flowers. But when I reach the Blue Border, then I can speak with decision; I know whereof I write, I know the variety and beauty of a garden bed of blue flowers. In blue you may have much difference in tint and quality without losing color effect. The Persian art workers have accomplished the combining of varying blues most wonderfully and successfully: purplish blues next to green-blues, and sapphire-blues alongside; and blues seldom clash in the flower beds.

Blue is my best beloved color; I love it as the bees love it. Every blue flower is mine; and I am as pleased as with a tribute of praise to a friend to learn that scientists have proved that blue flowers represent the most highly developed lines of descent. These learned men believe that all flowers were at first yellow, being perhaps only developed stamens; then some became white, others red; while the purple and blue were the latest and highest forms. The simplest shaped flowers, open to be visited by every insect, are still yellow or white, running into red or pink. Thus the Rose family have simple open symmetrical flowers; and there are no blue Roses — the flower has never risen to the blue stage. In the Pea family the simpler flowers are yellow or red; while the highly evolved members, such as Lupines,

Wistaria, Everlasting Pea, are purple or blue, varying to white. Bees are among the highest forms of insect life, and the labiate flowers are adapted to their visits; these nearly all have purple or blue petals—Thyme, Sage, Mint, Marjoram, Basil, Prunella, etc.

Of course the Blue Border runs into tints of pale lilac and purple and is thereby the gainer; but I

Scilla.

would remove from it the purple Clematis, Wistaria, and Passion-flower, all of which a friend has planted to cover the wall behind her blue flower bed. Sometimes the line between blue and purple is hard to define. Keats invented a word, *purplue*, which he used for this indeterminate color.

I would not, in my Blue Border, exclude an occasional group of flowers of other colors; I love a border of all colors far too well to do that. Here, as everywhere in my garden, should be white flowers, especially tall white flowers: white Foxgloves, white Delphinium, white Lupine, white Hollyhock, white Bell-flower, nor should I object to a few spires at one end of the bed of sulphur-yellow Lupines, or yellow Hollyhocks, or a group of Paris Daisies. I have seen a great Oriental Poppy growing in wonderful beauty near a mass of pale blue Larkspur, and Shirley Poppies are a delight with blues; and any one could arrange the pompadour tints of pink and blue in a garden who could in a gown.

Let me name some of the favorites of the Blue Border. The earliest but not the eldest is the pretty spicy Scilla in several varieties, and most satisfactory it is in perfection of tint, length of bloom, and great hardiness. It would be welcomed as we eagerly greet all the early spring blooms, even if it were not the perfect little blossom that is pictured on page 254, the very little Scilla that grew in my mother's garden.

The early spring blooming of the beloved Grape Hyacinth gives us an overflowing bowl of "blue principle"; the whole plant is imbued and fairly exudes blue. Ruskin gave the beautiful and appropriate term "blue-flushing" to this plant and others, which at the time of their blossoming send out through their veins their blue color into the surrounding leaves and the stem; he says they "breathe out" their color, and tells of a "saturated purple" tint.

Not content with the confines of the garden

border, the Grape Hyacinth has "escaped the garden," and become a field flower. The "seeing eye," ever quick to feel a difference in shade or

Sweet Alyssum Edging.

color, which often proves very slight upon close examination, viewed on Long Island a splendid sea of blue; and it seemed neither the time nor tint for the expected Violet. We found it a field of Grape Hyacinth, blue of leaf, of stem, of flower. While all flowers are in a sense perfect, some certainly do not appear so in shape, among the latter those of irregular sepals. Some flowers seem imperfect without any cause save the fancy of the one who is regarding them; thus to me the Balsam is an imperfect flower. Other flowers impress me delightfully with a sense of perfection. Such is the Grape Hyacinth, doubly grateful in this perfection in the time it comes in early spring. The Grape Hyacinth is the favorite spring flower of my garden — but no! I thought a minute ago the Scilla was! and what place has the Violet? the Flower de Luce? I cannot decide, but this I know — it is some blue flower.

Ruskin says of the Grape Hyacinth, as he saw it growing in southern France, its native home, "It was as if a cluster of grapes and a hive of honey had been distilled and pressed together into one small boss of celled and beaded blue." I always think of his term "beaded blue" when I look at it. There are several varieties, from a deep blue or purple to sky-blue, and one is fringed with the most delicate feathery petals. Some varieties have a faint perfume, and country folk call the flower "Baby's Breath" therefrom.

Purely blue, too, are some of our garden Hyacinths, especially a rather meagre single Hyacinth which looks a little chilly; and Gavin Douglas wrote in the springtime of 1500, "The Flower de Luce forth spread his heavenly bluc." It always jars upon my sense of appropriateness to hear this old garden favorite called Fleur de Lis. The accepted derivation of the word is that given by Grandmaison in his *Heraldic Dictionary*. Louis VII. of France, whose name was then written Loys, first gave the name to the flower, " Fleur de Loys " ; then it became Fleur de Louis, and finally, Fleur de Lis. Our flower caught its name from Louis. Tusser in

Bachelor's Buttons in a Salem Garden.

his list of flowers for windows and pots gave plainly Flower de Luce; and finally Gerarde called the plant Flower de Luce, and he advised its use as a domestic remedy in a manner which is in vogue in country homes in New England to-day. He said that the root " stamped plaister-wise, doth take away the blewnesse or blacknesse of any stroke" that is, a black and blue bruise. Another use advised of him is as obsolete as the form in which it was rendered. He said it was "good in a loch or licking medicine for shortness of breath." Our apothecaries no longer make, nor do our physicians prescribe, "licking medicines." The powdered root was urged as a complexion beautifier, especially to remove morphew, and as orris-root may be found in many of our modern skin lotions.

Ruskin most beautifully describes the Flower de Luce as the flower of chivalry — "with a sword for its leaf, and a Lily for its heart." These grand clumps of erect old soldiers, with leafy swords of green and splendid cuirasses and plumes of gold and bronze and blue, were planted a century ago in our grandmothers' garden, and were then Flower de Luce. A hundred years those sturdy sentinels have stood guard on either side of the garden gates — still Flower de Luce. There are the same clean-cut leaf swords, the same exquisite blossoms, far more beautiful than our tropical Orchids, though similar in shape; let us not change now their historic name, they still are Flower de Luce — the Flower de Louis.

The Violet family, with its Pansies and Ladies' Delights, has honored place in our Blue Border, though the rigid color list of a prosaic practical dyer finds these Violet allies a debased purple instead of blue.

Our wild Violets, the blue ones, have for me a sad lack for a Violet, that of perfume. They are not as lovely in the woodlands as their earlier coming neighbor, the shy, pure Hepatica. Bryant, calling the Hepatica Squirrelcups (a name I never heard given them elsewhere), says they form "a graceful company hiding in their bells a soft aerial blue." Of course, they vary through blue and pinky purple, but the blue is well hidden, and I never think of them save as an almost white flower. Nor are the Violets as lovely on the meadow and field slopes, as the mild Innocence, the Houstonia, called also Bluets, which is scarcely a distinctly blue expanse, but rather "a milky way of minute stars." An English botanist denies that it is blue at all. A field covered with Innocence always looks to me as if little clouds and puffs of blue-white smoke had descended and rested on the grass.

I well recall when the Aquilegia, under the name of California Columbine, entered my mother's garden, to which its sister, the red and yellow Columbine, had been brought from a rocky New England pasture when the garden was new. This Aquilegia came to us about the year 1870. I presume old catalogues of American florists would give details and dates of the journey of the plant from the Pacific to the Atlantic. It chanced that this first Aquilegia of my acquaintance was of a distinct light blue tint; and it grew apace and thrived and was vastly admired, and filled the border with blueness of that singular tint seen of late years in its fullest extent and most prominent position in the great masses of bloom of the blue Hydrangea, the show plant of such splendid summer homes as may be found at Newport. These blue Hydrangeas are ever to me a color blot. They accord with no other

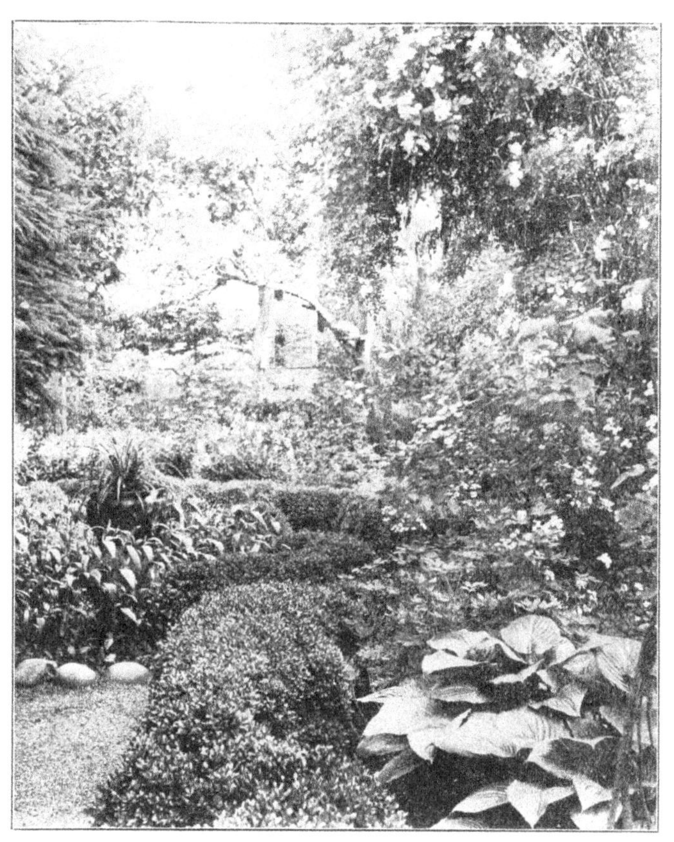

A "Sweet Garden-side" in Salem, Massachusetts.

flower and no foliage. I am ever reminded of blue mould, of stale damp. I looked with inexpressible aversion on a photograph of Cecil Rhodes' garden at Cape Town — several solid acres set with this blue Hydrangea and nothing else, unbroken by tree or shrub, and scarce a path, growing as thick as a field sown with ensilage corn, and then I thought what would be the color of that mass! that crop of Hydrangeas! Yet I am told that Rhodes is a flower-lover and flower-thinker. Now this Aquilegia was of similar tint; it was blue, but it was not a pleasing blue, and additional plants of pink, lilac, and purple tints had to be added before the Aquilegia was really included in our list of well-beloveds.

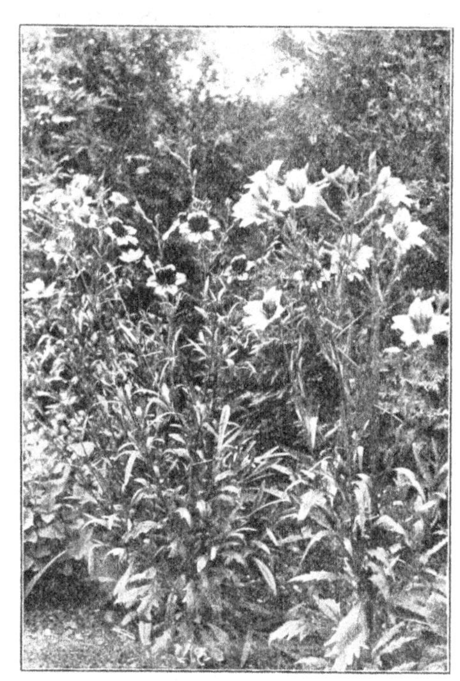

Salpiglossis.

There are other flowers for the blue border. It

is pleasant to plant common Flax, if you have ample room; it is a superb blue; to many persons the blossom is unfamiliar, and is always of interest. Its lovely flowers have been much sung in English verse. The Salpiglossis, shown on the opposite page, is in its azure tint a lovely flower, though it is a kinsman of the despised Petunia.

How the Campanulaceæ enriched the beauty and the blueness of the garden. We had our splendid clusters of Canterbury Bells, both blue and white I have told elsewhere of our love for them in childhood. Equally dear to us was a hardy old Campanula whose full name I know not, perhaps it is the Pyramidalis; it is shown on page 263, the very plant my mother set out, still growing and blooming; nothing in the garden is more gladly welcomed from year to year. It partakes of the charm shared by every bell-shaped flower, a simple form, but an ever pleasing one. We had also the *Campanula persicifolia* and *trachelium*, and one we called Bluebells of Scotland, which was not the correct name. It now has died out, and no one recalls enough of its exact detail to learn its real name. The showiest bell-flower was the *Platycodon grandiflorum*, the Chinese or Japanese Bell-flower, shown on page 264. Another name is the Balloon-flower, this on account of the characteristic buds shaped like an inflated balloon. It is a lovely blue in tint, though this photograph was taken from a white-flowered plant in the white border at Indian Hill. The Giant Bell-flower is a *fin de siècle* blossom named *Ostrowskia*, with flowers four inches deep and six inches in diameter;

it has not yet become common in our gardens, where the *Platycodon* rules in size among its bell-shaped fellows.

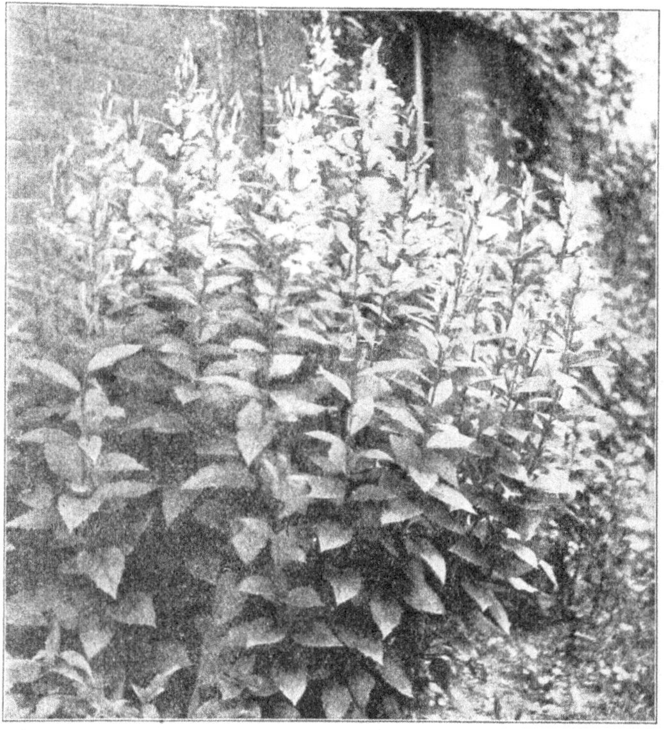

The Old Campanula.

There are several pretty low-growing blue flowers suitable for edgings, among them the tiny stars of the Swan River Daisy (*Brachycome iberidifolia*) sold as purple, but as brightly blue as Scilla. The dwarf Ageratum is also a long-blossoming soft-tinted blue flower; it made a charming edging in my sister's garden last summer; but I should never put either of them on the edge of the blue border.

Chinese Bell-flower.

The dull blue, sparsely set flowers of the various members of the Mint family have no beauty in color, nor any noticeable elegance; the Blue Sage is the only vivid-hued one, and it is a true ornament to the border. Prunella was ever found in old gardens, now it is a wayside weed. Thoreau loved the Prunella for its blueness, its various lights, and noted that its color deepened toward night. This flower, regarded with indifference by nearly every one, and distaste by many, always to him suggested coolness and freshness by its presence. The Prunella was beloved also by

Ruskin, who called it the soft warm-scented Brunelle, and told of the fine purple gleam of its hooded blossom: "the two uppermost petals joined like an old-fashioned enormous hood or bonnet; the lower petal torn deep at the edges into a kind of fringe," — and he said it was a "Brownie flower," a little eerie and elusive in its meaning. I do not like it because it has such a disorderly, unkempt look, it always seems bedraggled.

The pretty ladder-like leaf of Jacob's Ladder is most delicate and pleasing in the garden, and its blue bell-flowers are equally refined. This is truly an old-fashioned plant, but well worth universal cultivation.

In answer to the question, What is the bluest flower in the garden or field? one answered Fringed Gentian; another the Forget-me-not, which has much pink in its buds and yellow in its blossoms; another Bee Larkspur; and the others *Centaurea cyanus* or Bachelor's Buttons, a local American name for them, which is not even a standard folk name, since there are twenty-one English plants called Bachelor's Buttons. Ragged Sailor is another American name. Corn-flower, Blue-tops, Blue Bonnets, Bluebottles, Loggerheads are old English names. Queerer still is the title Break-your-spectacles. Hawdods is the oldest name of all. Fitzherbert, in his *Boke of Husbandry*, 1586, thus describes briefly the plant: —

"Hawdod hath a blewe floure, and a few lytle leaves, and hath fyve or syxe branches floured at the top."

In varied shades of blue, purple, lilac, pink, and white, Bachelor's Buttons are found in every old garden, growing in a confused tangle of "lytle leaves" and vari-colored flowers, very happily and with very good effect. The illustration on page 258 shows their growth and value in the garden.

In *The Promise of May* Dora's eyes are said to be as blue as the Bluebell, Harebell, Speedwell, Bluebottle, Succory, Forget-me-not, and Violets; so we know what flowers Tennyson deemed blue.

Another poet named as the bluest flower, the Monk's-hood, so wonderful of color, one of the very rarest of garden tints; graceful of growth, blooming till frost, and one of the garden's delights. In a list of garden flowers published in Boston, in 1828, it is called Cupid's Car. Southey says in *The Doctor*, of Miss Allison's garden: "The Monk's-hood of stately growth Betsey called 'Dumbledores Delight,' and was not aware that the plant, in whose helmet- rather than cowl-shaped flowers, that busy and best-natured of all insects appears to revel more than any other, is the deadly Aconite of which she read in poetry." The dumbledore was the bumblebee, and this folk name was given, as many others have been, from a close observance of plant habits; for the fertilization of the Monk's-hood is accomplished only by the aid of the bumblebee.

Many call Chicory or Succory our bluest flower. Thoreau happily termed it "a cool blue." It is not often the fortune of a flower to be brought to notice and affection because of a poem; we expect the poem to celebrate the virtues of flowers already

Garden at Tudor Place.

loved. The Succory is an example of a plant, known certainly to flower students, yet little thought of by careless observers until the beautiful poem of Margaret Deland touched all who read it. I think this a gem of modern poesy, having in full that great element of a true poem, the most essential element indeed of a short poem — the power of suggestion. Who can read it without being stirred by its tenderness and sentiment, yet how few are the words.

> "Oh, not in ladies' gardens,
> My peasant posy,
> Shine thy dear blue eyes;
> Nor only — nearer to the skies
> In upland pastures, dim and sweet,
> But by the dusty road,
> Where tired feet
> Toil to and fro,
> Where flaunting Sin
> May see thy heavenly hue,
> Or weary Sorrow look from thee
> Toward a tenderer blue."

I recall perfectly every flower I saw in pasture, swamp, forest, or lane when I was a child; and I know I never saw Chicory save in old gardens. It has increased and spread wonderfully along the roadside within twenty years. By tradition it was first brought to us from England by Governor Bowdoin more than a century ago, to plant as forage.

In our common Larkspur, the old-fashioned garden found its most constant and reliable blue banner, its most valuable color giver. Self-sown, this Larkspur sprung up freely every year; needing no special cherishing or nourishing, it grew apace, and bloomed with a luxuriance and length of flowering that cheerfully blued the garden for the whole summer. It was a favorite of children in their floral games, and pretty in the housewife's vases, but its chief hold on favor was in its democracy and endurance. Other flowers drew admirers and lost them; some grew very ugly in their decay; certain choice seedlings often had stunted development, garden scourges attacked tender beauties; fierce July suns dried up the whole border, all save the Larkspur, which neither withered nor decayed; and often, unaided, saved the midsummer garden from scanty unkemptness and dire disrepute.

The graceful line of Dr. Holmes, "light as a loop of Larkspur," always comes to my mind as I look at a bed of Larkspur; and I am glad to show here a "loop of Larkspur," growing by the great boulder which he loved in the grounds of his country home at Beverly Farms. I liked to fancy that Dr. Holmes's expression was written by him from his memory of the little wreaths and garlands of pressed Larkspur that have been made so universally for over a century by New England children. But that careful flower observer, Mrs. Wright, notes that in a profuse growth of the Bee Larkspur, the strong flower spikes often are in complete loops before full expansion into a straight spire; some are looped thrice. Dr. Holmes was a minute observer of floral characteristics, as is shown in his poem on the

"Light as a Loop of Larkspur."

Coming of Spring, and doubtless saw this curious growth of the Larkspur.

Common annual Larkspurs now are planted

in every one's garden, and deservedly grow in favor yearly. The season of their flowering can be prolonged, renewed in fact, by cutting away the withered flower stems. They respond well to all caretaking, to liberal fertilizing and watering, just as they dwindle miserably with neglect. There are a hundred varieties in all; among them the "Rocket-flowered" and "Ranunculus flowered" Larkspurs or Delphiniums are ever favorites. A friend burst forth in railing at being asked to admire a bed of Delphinium. "Why can't she call them the good old-time name of Larkspur, and not a stiff name cooked up by the botanists." I answered naught, but I remembered that Parkinson in his *Garden of Pleasant Flowers* gives a chapter to Delphinium, with Lark's-heel as a second thought. "Their most usual name with us," he states, "is Delphinium." There is meaning in the name: the flower is dolphin-like in shape. Of the perennial varieties the *Delphinium brunonianum* has lovely clear blue, musk-scented flowers; the Chinese or Branching Larkspur is of varied blue tints and tall growth, and blooms from midsummer until frost. And loveliest of all, an old garden favorite, the purely blue Bee Larkspur, with a bee in the heart of each blossom. In an ancient garden in Deerfield I saw this year a splendid group of plants of the old *Delphinium Belladonna:* it is a weak-kneed, weak-backed thing; but give it unobtrusive crutches and busks and backboards (in their garden equivalents), and its incomparable blue will reward your care. There is something singular in the blue of Larkspur. Even on a dark night you can see it showing a distinct blue in the garden like a blue lambent flame.

> "Larkspur lifting turquoise spires
> Bluer than the sorcerer's fires."

Mrs. Milne-Home says her old Scotch gardener called the white Delphinium Elijah's Chariot — a resounding, stately title. Helmet-flower is another name. I think the Larkspur Border, and the Blue Border both gain if a few plants of the pure white Delphinium, especially the variety called the Emperor, bloom by the blue flowers. In our garden the common blue Larkspur loves to blossom by the side of the white Phlox. A bit of the border is shown on page 162. In another corner of the garden the pink and lilac Larkspur should be grown; for their tints, running into blue, are as varied as those of an opal.

I have never seen the wild Larkspur which grows so plentifully in our middle Southern states; but I have seen expanses of our common garden Larkspur which has run wild. Nor have I seen the glorious fields of Wyoming Larkspur, so poisonous to cattle; nor the magnificent Larkspur, eight feet high, described so radiantly to us by John Muir, which blues those wonders of nature, the hanging meadow gardens of California.

I am inclined to believe that Lobelia is the least pleasing blue flower that blossoms. I never see it in any place or juxtaposition that it satisfies me. When you take a single flower of it in your hand, its single little delicate bloom is really just as pretty as Blue-eyed Grass, or Innocence, or Scilla, and the whole plant regarded closely by itself isn't at all bad; but whenever and wherever you find it growing in a garden, you never want it in *that* place, and you shift it here and there. I am convinced that the Lobelia is simply impossible; it is an alien, wrong in some subtle way in tint, in habit of growth, in time of blooming. The last time I noted it in any large garden planting, it was set around the roots of some standard Rose bushes; and the gardener had displayed some thought about it; it was only at the base of white or cream-yellow Roses; but it still was objectionable. I think I would exterminate Lobelia if I could, banish it and forget it. In the minds of many would linger a memory of certain ornate garden vases, each crowded with a Pandanus-y plant, a pink Begonia, a scarlet double Geranium, a purple Verbena or a crimson Petunia, all gracefully entwined with Nasturtiums and Lobelia — while these folks lived, the Lobelia would not be forgotten.

You will have some curious experiences with your Blue Border; kindly friends, pleased with its beauty or novelty, will send to you plants and seeds to add to its variety of form "another bright blue flower." You will usually find you have added variety of tint as well, ranging into crimson and deep purple, for color blindness is far more general than is thought.

The loveliest blue flowers are the wild ones of fields and meadows; therefore the poor, says Alphonse Karr, with these and the blue of the sky have the best and the most of all blueness. Yet we are constantly hearing folks speak of the lack of the color blue among wild flowers, which always surprises me; I suppose I see blue because I love blue. In pure cobalt tint it is rare; in compensation, when it does abound, it makes a permanent imprint on our vision, which never vanishes. Recalling in midwinter the expanses of color in summer waysides, I do not see them white with Daisies, or yellow with Goldenrod, but they are in my mind's vision brightly, beautifully blue. One special scene is the blue of Fringed Gentians, on a sunny October day, on a rocky hill road in Royalston, Massachusetts, where they sprung up, wide open, a solid mass of blue, from stone wall to stone wall, with scarcely a wheel rut showing among them. Even thus, growing in as lavish abundance as any weed, the Fringed Gentian still preserved in collective expanse, its delicate, its distinctly aristocratic bearing.

Bryant asserts of this flower: —

> "Thou waitest late, and com'st alone
> When woods are bare, and birds are flown."

But by this roadside the woods were far from bare. Many Asters, especially the variety I call Michaelmas Daisies, Goldenrod, Butter-and-eggs, Turtle Head, and other flowers, were in ample bloom. And the same conditions of varied flower companionship existed when I saw the Fringed Gentian blooming near Bryant's own home at Cummington.

Another vast field of blue, ever living in my memory, was that of the Viper's Bugloss, which I viewed with surprise and delight from the platform of a train, returning from the Columbian Exposition; when I asked a friendly brakeman what the flower was called, he answered "Vilets," as nearly all workingmen confidently name every blue flower; and he sprang from the train while the locomotive was swallowing water, and brought to me a great armful of blueness. I am not wont to like new flowers as well as my childhood's friends, but I found this new friend, the Viper's Bugloss, a very welcome and pleasing acquaintance. Curious, too, it is, with the red anthers exserted beyond the bright blue corolla, giving the field, when the wind blew across it, a new aspect and tint, something like a red and blue changeable silk. The Viper's Bugloss seems to have the perva-

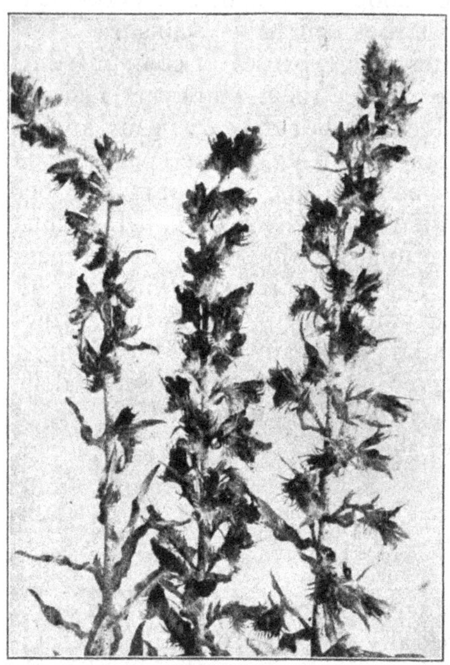

Viper's Bugloss.

sive power of many another blue and purple flower, Lupine, Iris, Innocence, Grape Hyacinth, Vervain, Aster, Spiked Loosestrife; it has become in many states a tiresome weed. On the Esopus Creek (which runs into the Hudson River) and adown the Hudson, acre after acre of meadow and field by the waterside are vivid with its changeable hues, and the New York farmers' fields are overrun by the newcomer.

I have seen the Viper's Bugloss often since that day on the railroad train, now that I know it, and think of it. Thoreau noted the fact that in a large sense we find only what we look for. And he defined well our powers of perception when he said that many an object will not be seen, even when it comes within the range of our visual ray, because it does not come within the range of our intellectual ray.

Last spring, having to spend a tiresome day riding the length of Long Island, I beguiled the hours by taking with me Thoreau's *Summer* to compare his notes of blossomings with those we passed. It was June 5, and I read:—

"The Lupine is now in its glory. It is the more important because it occurs in such extensive patches, even an acre or more together. . . . It paints a whole hillside with its blue, making such a field, if not a meadow, as Proserpine might have wandered in. Its leaf was made to be covered with dewdrops. I am quite excited by this prospect of blue flowers in clumps, with narrow intervals; such a profusion of the heavenly, the Elysian color, as if these were the Elysian Fields. That is the value of the Lupine. The earth is blued with it. . . . You may have passed here a fortnight ago and the field was comparatively barren. Now you come, and these glorious redeemers appear to have flashed out here all at once. Who plants the seeds of Lupines in the barren soil? Who watereth the Lupines in the field?"

The Precision of Leaf and Flower of Lupine.

I looked from a car window, and lo! the Long Island Railroad ran also through an Elysian Field of Lupines, nay, we sailed a swift course through a summer sea of blueness, and I seem to see it still, with its prim precision of outline and growth of both leaf and flower. The Lupine is beautiful in the garden border as it is in the landscape, whether the blossom be blue, yellow, or white.

Thoreau was the slave of color, but he was the master of its description. He was as sensitive as Keats to the charm of blue, and left many records of his love, such as the paragraphs above quoted.

He noted with delight the abundance of "that principle which gives the air its azure color, which makes the distant hills and meadows appear blue," the "great blue presence" of Monadnock and Wachusett with its "far blue eye." He loved Lowell's

> "Sweet atmosphere of hazy blue,
> So leisurely, so soothing, so forgiving,
> That sometimes makes New England fit for living."

He revelled in the blue tints of water, of snow, of ice; in "the blueness and softness of a mild winter day." The constant blueness of the sky at night thrilled him with "an everlasting surprise," as did the blue shadows within the woods and the blueness of distant woods. How he would have rejoiced in Monet's paintings, how true he would have found their tones. He even idealized blueberries, "a very innocent ambrosial taste, as if made of ether itself, as they are colored with it."

Thoreau was ever ready in thought of Proserpina gathering flowers. He offers to her the Lupine, the Blue-eyed Grass, and the Tufted Vetch, "blue, inclining in spots to purple"; it affected him deeply to see such an abundance of blueness in the grass. "Celestial color, I see it afar in masses on the hillside near the meadow — so much blue."

I usually join with Thoreau in his flower loves; but I cannot understand his feeling toward the blue Flag; that, after noting the rich fringed recurved parasols over its anthers, and its exquisite petals, that he could say it is "a little too showy and gaudy, like some women's bonnets." I note that whenever he compares flowers to women it is in no flattering humor to either; which is, perhaps, what we expect from a man who chose to be a bachelor and a hermit. His love of obscure and small flowers might explain his sentiment toward the radiant and dominant blue Flag.

The most valued flower of my childhood, outside the garden, was a little sister of the Iris — the Blue-eyed Grass. To find it blooming was a triumph, for it was not very profuse of growth near my home; to gather it a delight; why, I know not, since the tiny blooms promptly closed and withered as soon as we held them in our warm little hands. Colonel Higginson writes wittily of the Blue-eyed Grass, "It has such an annoying way of shutting up its azure orbs the moment you gather it; and you reach home with a bare stiff blade which deserves no better name than *Sisyrinchium anceps*."

The only time I ever played truant was to run off one June morning to find "the starlike gleam amid the grass and dew"; to pick Blue-eyed Grass in a field to which I was conducted by another naughty girl. I was simple enough to come home at mid-day with my hands full of the stiff blades and tightly closed blooms; and at my mother's inquiry as to my acquisition of these treasures, I promptly burst into tears. I was then told, in impressive phraseology adapted to my youthful comprehension, and with the flowers as eloquent proof, that all stolen pleasures were ever like my coveted flowers, withered and unsightly as soon as gathered — which my mother believed was true.

The blossoms of this little Iris seem to lie on the surface of the grass like a froth of blueness; they gaze up at the sky with a sort of intimacy as if they were a part of it. Thoreau called it an "air of easy sympathy." The slightest clouding or grayness of atmosphere makes them turn away and close.

The naming of Proserpina leads me to say this: that to grow in love and knowledge of flowers, and above all of blue flowers, you must read Ruskin's *Proserpina.* It is a book of botany, of studies of plants, but begemmed with beautiful sentences and thoughts and expressions, with lessons of pleasantness which you can never forget, of pictures which you never cease to see, such sentences and pictures as this: —

> "Rome. My father's Birthday. I found the loveliest blue Asphodel I ever saw in my life in the fields beyond Monte Mario — a spire two feet high, of more than two hundred stars, the stalks of them all deep blue as well as the flowers. Heaven send all honest people the gathering of the like, in the Elysian Fields, some day!"

Oh, the power of written words! when by these few lines I can carry forever in my inner vision this spire of starry blueness. To that writer, now in the Elysian Fields, an honest teacher if ever one lived, I send my thanks for this beautiful vision of blueness.

CHAPTER XII

PLANT NAMES

"The fascination of plant names is founded on two instincts,—love of Nature and curiosity about Language."

— *English Plant Names*, Rev. John Earle, 1880.

ERBAL magic is the subtle mysterious power of certain words. This power may come from association with the senses; thus I have distinct sense of stimulation in the word scarlet, and pleasure in the words lucid and liquid. The word garden is a never ceasing delight; it seems to me Oriental; perhaps I have a transmitted sense from my grandmother Eve of the Garden of Eden. I like the words, a Garden of Olives, a Garden of Herbs, the Garden of the Gods, a Garden enclosed, Philosophers of the Garden, the Garden of the Lord. As I have written on gardens, and thought on gardens, and walked in gardens, "the very music of the name has gone into my being." How beautiful are Cardinal Newman's words: —

"By a garden is meant mystically a place of spiritual repose, stillness, peace, refreshment, delight."

There was, in Gerarde's day, no fixed botanical nomenclature of any of the parts or attributes of a

The Garden's Friend.

plant. Without using botanical terms, try to describe a plant so as to give an exact notion of it to a person who has never seen it, then try to find common words to describe hundreds of plants; you will then admire the vocabulary of the old herbalist, his "fresh English words," for you will find that it needs the most dextrous use of words to convey accurately the figure of a flower. That felicity and facility Gerarde had; "a bleak white color" — how clearly you see it! The Water Lily had "great round leaves like a buckler." The Cat-tail Flags "flower and bear their mace or torch in July and August." One plant had "deeply gashed leaves." The Marigold had "fat thick crumpled leaves set upon a gross and spongious stalke." Here is the Wake-robin, "a long hood in proportion like the ear of a hare, in middle of which hood cometh forth a pestle or clapper of a dark murry or pale purple color." The leaves of the Corn-marigold are "much hackt and cut into divers sections and placed confusedly." Another plant had leaves of "an overworne green," and Pansy leaves were "a bleak green." The leaves of Tansy are also vividly described as "infinitely jagged and nicked and curled with all like unto a plume of feathers."

The classification and naming of flowers was much thought and written upon from Gerarde's day, until the great work of Linnæus was finished. Some very original schemes were devised. *The Curious and Profitable Gardner*, printed in 1730, suggested this plan: That all plants should be named to indicate their color, and that the initials of their names should be the initials of their respective colors; thus if a plant were named William the Conqueror it would indicate that the name was of a white flower with crimson lines or shades. "Virtuous Oreada would indicate a violet and orange flower; Charming Phyllis or Curious Plotinus a crimson and purple blossom." S. was to indicate

Black or Sable, and what letter was Scarlet to have? The "curious ingenious Gentleman" who published this plan urged also the giving of "pompous names" as more dignified; and he made the assertion that French and Flemish "Flowerists" had adopted his system.

Edging of Striped Lilies in a Salem Garden.

These were all forerunners of Ruskin, with his poetical notions of plant nomenclature, such as this; that feminine forms of names ending in *a* (as Prunella, Campanula, Salvia, Kalmia) and *is* (Iris, Amarylis) should be given only to plants "that are pretty and good"; and that real names, Lucia, Clarissa, etc., be also given. Masculine names in *us* should be given to plants of masculine qualities, — strength, force, stubbornness; neuter endings in *um*, given to plants indicative of evil or death.

I have a fancy anent many old-time flower names that they are also the names of persons. I think of them as persons bearing various traits and characteristics. On the other hand, many old English Christian names seem so suited for flowers, that they might as well stand for flowers as for persons. Here are a few of these quaint old names, Collet, Colin, Emmot, Issot, Doucet, Dobinet, Cicely, Audrey, Amice, Hilary, Bryde, Morrice, Tyffany, Amery, Nowell, Ellice, Digory, Avery, Audley, Jacomin, Gillian, Petronille, Gresel, Joyce, Lettice, Cibell, Avice, Cesselot, Parnell, Renelsha. Do they not "smell sweet to the ear"? The names of flowers are often given as Christian names. Children have been christened by the names Dahlia, Clover, Hyacinth, Asphodel, Verbena, Mignonette, Pansy, Heartsease, Daisy, Zinnia, Fraxinella, Poppy, Daffodil, Hawthorn.

What power have the old English names of garden flowers, to unlock old memories, as have the flowers themselves! Dr. Earle writes, "The fascination of plant names is founded on two instincts; love of Nature, and curiosity about Language." To these I should add an equally strong instinct in many persons — their sensitiveness to associations.

I am never more filled with a sense of the delight of old English plant-names than when I read the liquid verse of Spenser: —

"Bring hether the pincke and purple Cullembine
. . . with Gellifloures,
Bring hether Coronations and Sops-in-wine
Worne of paramours.
Sow me the ground with Daffadowndillies
And Cowslips and Kingcups and loved Lilies,
The pretty Pawnce
The Chevisaunce
Shall match with the fayre Flour Delice."

Why, the names are a pleasure, though you know not what the Sops-in-wine or the Chevisaunce were. Gilliflowers were in the verses of every poet. One of scant fame, named Plat, thus sings:—

"Here spring the goodly Gelofors,
Some white, some red in showe;
Here pretie Pinks with jagged leaves
On rugged rootes do growe;
The Johns so sweete in showe and smell,
Distinct by colours twaine,
About the borders of their beds
In seemlie sight remaine."

If there ever existed any difference between Sweet-johns and Sweet-williams, it is forgotten now. They have not shared a revival of popularity with other old-time favorites. They were one of the "garland flowers" of Gerarde's day, and were "esteemed for beauty, to deck up the bosoms of the beautiful, and for garlands and crowns of pleasure." In the gardens of Hampton Court in the days of King Henry VIII., were Sweet-williams, for the plants had been bought by the bushel. Sweet-williams are little sung by the poets, and I never knew any one to call the Sweet-william her favorite flower, save one person. Old residents of Worcester will recall the tiny cottage that stood on the corner of Chestnut and Pleasant streets, since the remote years when the latter-named street was a post-road. It was occupied during my childhood by friends of my mother — a century-old mother, and her ancient unmarried daughter. Behind the house stretched one of the most cheerful gardens I have ever seen; ever, in my memory, bathed in glowing sunlight and color. Of its glories I recall specially the long spires of vivid Bee Larkspur, the varied Poppies of wonderful growth, and the rioting Sweet-williams. The latter flowers had some sentimental association to the older lady, who always asserted with emphasis to all visitors that they were her favorite flower. They overran the entire garden, crowding the grass plot where the washed garments were hung out to dry, even growing in the chinks of the stone steps and between the flat stone flagging of the little back yard, where stood the old well with its moss-covered bucket. They spread under the high board fence and appeared outside on Chestnut Street; and they extended under the dense Lilac bushes and Cedars and down the steep grass bank and narrow steps to Pleasant Street. The seed was carefully gathered, especially of one glowing crimson beauty, the color of the Mullein Pink, and a gift of it was highly esteemed by other garden owners. Old herbals say the Sweet-williams are "worthy the Respect of the Greatest Ladies who are Lovers of Flowers." They

Garden Seat at Avonwood Court.

certainly had the respect and love of these two old ladies, who were truly Lovers of Flowers.

I recall an objection made to Sweet-williams, by some one years ago, that they were of no use or value save in the garden; that they could never be combined in bouquets, nor did they arrange well in vases. It is a place of honor, some of us believe, to be a garden flower as well as a vase flower. This garden was the only one I knew when a child which contained plants of Love-lies-bleeding — it had even then been deemed old-fashioned and out of date. And it also held a few Sunflowers, which had not then had a revival of attention, and seemed as obsolete as the Love-lies-bleeding. The last-named flower I always disliked, a shapeless, gawky creature, described in florists' catalogues and like publications as "an effective plant easily attaining to a splendid form bearing many plume-tufts of rich lustrous crimson." It is the "immortal amarant" chosen by Milton to crown the celestial beings in *Paradise Lost*. Poor angels! they have had many trying vagaries of attire assigned to them.

I can contribute to plant lore one fantastic notion in regard to Love-lies-bleeding — though I can find no one who can confirm this memory of my childhood. I recall distinctly expressions of surprise and regret that these two old people in Worcester should retain the Love-lies-bleeding in their garden, because "the house would surely be struck with lightning." Perhaps this fancy contributed to the exile of the flower from gardens.

There be those who write, and I suppose they

believe, that a love of Nature and perception of her beauties and a knowledge of flowers, are the dower of those who are country born and bred; by which is meant reared upon a farm. I have not found this true. Farm children have little love for Nature and are surprisingly ignorant about wild flowers, save a

Terraced Garden of the Misses Nichols, Salem, Massachusetts.

very few varieties. The child who is garden bred has a happier start in life, a greater love and knowledge of Nature. It is a principle of Froebel that one must limit a child's view in order to coördinate his perceptions. That is precisely what is done in a child's regard of Nature by his life in a garden; his view is limited and he learns to know garden flowers and birds and insects thoroughly, when the vast and bewildering variety of field and forest would have remained unappreciated by him.

It is a distressing condition of the education of farmers, that they know so little about the country. The man knows about his crops, and his wife about the flowers, herbs, and vegetables of her garden; but no countrymen know the names of wild flowers — and few countrywomen, save of medicinal herbs. I asked one farmer the name of a brilliant autumnal flower whose intense purple was then unfamiliar to me — the Devil's-bit. He answered, "Them's Woilets." Violet is the only word in which the initial V is ever changed to W by native New Englanders. Every pink or crimson flower is a Pink. Spring blossoms are "Mayflowers." A frequent answer is, "Those ain't flowers, they're weeds." They are more knowing as to trees, though shaky about the evergreen trees, having little idea of varieties and inclined to call many Spruce. They know little about the reasons for names of localities, or of any historical traditions save those of the Revolution. One exclaims in despair, "No one in the country knows anything about the country."

This is no recent indifference and ignorance; Susan Cooper wrote in her *Rural Hours* in 1848: —

"When we first made acquaintance with the flowers of the neighborhood we asked grown persons — learned perhaps in many matters — the common names of plants they must have seen all their lives, and we found they were no wiser than the children or ourselves. It is really surprising how little country people know on such subjects. Farmers and their wives can tell you nothing on these matters. The men are at fault even among the trees on their own farms, if they are at all out of the common way; and as for smaller native plants, they know less about them than Buck or Brindle, their own oxen."

Kitchen Dooryard at Wilbour Farm, Kingston, Rhode Island.

In that delightful book, *The Rescue of an Old Place*, the author has a chapter on the love of flowers in America. It was written anent the ever-present statements seen in metropolitan print that Americans do not love flowers because they are used among the rich and fashionable in large cities for extravagant display rather than for enjoyment; and that we accept botanical names for our indigenous plants instead of calling them by homely ones such as familiar flowers are known by in older lands.

Two more foolish claims could scarcely be made. In the first place, the doings of fashionable folk in large cities are fortunately far from being a national index or habit. Secondly, in ancient lands the people named the flowers long before there were botanists, here the botanists found the flowers and named them for the people. Moreover, country folk in New England and even in the far West call flowers by pretty folk-names, if they call them at all, just as in Old England.

The fussing over the use of the scientific Latin names for plants apparently will never cease; many of these Latin names are very pleasant, have become so from constant usage, and scarcely seem Latin; thus Clematis, Tiarella, Rhodora, Arethusa, Campanula, Potentilla, Hepatica. When I know the folk-names of flowers I always speak thus of them — and *to them;* but I am grateful too for the scientific classification and naming, as a means of accurate distinction. For any flower student quickly learns that the same English folk-name is given in different localities to very different plants. For instance, the name Whiteweed is applied to ten different plants; there are in England ten or twelve Cuckoo-flowers, and twenty-one Bachelor's Buttons. Such names as Mayflower, Wild Pink, Wild Lily, Eyebright, Toad-flax, Ragged Robin, None-so-pretty, Lady's-fingers, Four-o'clocks, Redweed, Buttercups, Butter-flower, Cat's-tail, Rocket, Blue-Caps, Creeping-jenny, Bird's-eye, Bluebells, apply to half a dozen plants.

The old folk-names are not definite, but they are delightful; they tell of mythology and medicine, of superstitions and traditions; they show trains of relationship, and associations; in fact, they appeal more to the philologist and antiquarian than to the botanist. Among all the languages which contribute to the variety and picturesqueness of English plant

"A running ribbon of perfumed snow which the sun is melting rapidly.

names, Dr. Prior deems Maple the only one surviving from the Celtic language. Gromwell and Wormwood may possibly be added.

There are some Anglo-Saxon words; among them Hawthorn and Groundsel. French, Dutch, and Danish names are many, Arabic and Persian are more. Many plant names are dedicatory; they embody the names of the saints and a few the names of the Deity. Our Lady's Flowers are many and interesting; my daughter wrote a series of articles for the *New York Evening Post* on Our Lady's Flowers, and the list swelled to a surprising number. The devil and witches have their shares of flowers, as have the fairies.

I have always regretted deeply that our botanists neglected an opportunity of great enrichment in plant nomenclature when they ignored the Indian names of our native plants, shrubs, and trees. The first names given these plants were not always planned by botanists; they were more often invented in loving memory of English plants, or sometimes from a fancied resemblance to those plants. They did give the wonderfully descriptive name of Moccasin-flower to that creature of the wild-woods; and a far more appropriate title it is than Lady's-slipper, but it is not as well known. I have never found the Lady's-slipper as beautiful a flower as do nearly all my friends, as did my father and mother, and I was pleased at Ruskin's sharp comment that such a slipper was only fit for very gouty old toes.

Pappoose-root utilizes another Indian word. Very few Indian plant names were adopted by the white men, fewer still have been adopted by the scientists. The *Catalpa speciosa* (Catalpa); the *Zea mays* (Maize); and *Yucca filamentosa* (Yucca), are the only ones I know. Chinkapin, Cohosh, Hackmatack, Kinnikinnik, Tamarack, Persimmon, Tupelo, Squash, Puccoon, Pipsissewa, Musquash, Pecan, the Scuppernong and Catawba grapes, are our only well-known Indian plant names that survive. Of these Maize, the distinctive product of the United States, will ever link us with the vanishing Indian. It will be noticed that only Puccoon, Cohosh, Pipsissewa, Hackmatack, and Yucca are names of flowering plants; of these Yucca is the only one generally known. I am glad our stately native trees, Tupelo, Hickory, Catalpa, bear Indian names.

A curious example of persistence, when so much else has perished, is found in the word "Kiskatomas," the shellbark nut. This Algonquin word was heard everywhere in the state of New York sixty years ago, and is not yet obsolete in families of Dutch descent who still care for the nut itself.

We could very well have preserved many Indian names, among them Hiawatha's

> "Beauty of the springtime,
> The Miskodeed in blossom,"

I think Miskodeed a better name than Claytonia or Spring Beauty. The Onondaga Indians had a suggestive name for the Marsh Marigold, "It-opens-the-swamps," which seems to show you the yellow stars "shining in swamps and hollows gray." The name Cowslip has been transferred to it in some localities in New England, which is not strange when we find that the flower has fifty-six English folk-names; among them are Drunkards, Crazy Bet, Meadow-bright, Publicans and Sinners, Soldiers' Buttons, Gowans, Kingcups, and Buttercups. Our Italian street venders call them Buttercups. In erudite Boston, in sight of Boston Common, the beautiful Fringed Gentian is not only called, but

Fountain Garden at Sylvester Manor.

labelled, French Gentian. To hear a lovely bunch of the Arethusa called Swamp Pink is not so strange. The Sabbatia grows in its greatest profusion in the vicinity of Plymouth, Massachusetts, and is called locally, "The Rose of Plymouth." It is sold during its season of bloom in the streets of that town and is used to dress the churches. Its name was given to honor an early botanist, Tiberatus Sabbatia, but in Plymouth there is an almost universal belief that it was named because the Pilgrims of 1620 first saw the flower on the Sabbath day. It thus is regarded as a religious emblem, and strong objection is made to mingling other flowers with it in church decoration. This legend was invented about thirty years ago by a man whose name is still remembered as well as his work.

CHAPTER XIII

TUSSY-MUSSIES

"There be some flowers make a delicious Tussie-Mussie or Nosegay both for Sight and Smell."

—John Parkinson, *A Garden of all Sorts of Pleasant Flowers*, 1629.

O following can be more productive of a study and love of word derivations and allied word meanings than gardening. An interest in flowers and in our English tongue go hand in hand. The old mediæval word at the head of this chapter has a full explanation by Nares as "A nosegay, a tuzzie-muzzie, a sweet posie." The old English form, *tussy-mose* was allied with *tosty*, a bouquet, *tuss* and *tusk*, a wisp, as of hay, *tussock*, and *tutty*, a nosegay. Thomas Campion wrote:—

"Joan can call by name her cows,
And deck her windows with green boughs;
She can wreathes and tuttyes make,
And trim with plums a bridal cake."

Tussy-mussy was not a colloquial word; it was found in serious, even in religious, text. A tussy-mussy was the most beloved of nosegays, and was often made of flowers mingled with sweet-scented leaves.

My favorite tussy-mussy, if made of flowers, would be of Wood Violet, Cabbage Rose, and Clove Pink. These are all beautiful flowers, but many of our most delightful fragrances do not come from flowers of gay dress; even these three are not showy flowers; flowers of bold color and growth are not apt to be sweet-scented; and all flower perfumes of great distinction, all that are unique, are from blossoms of modest color and bearing. The Calycanthus, called Virginia Allspice, Sweet Shrub, or Strawberry bush, has what I term a perfume of distinction, and its flowers are neither fine in shape, color, nor quality.

I have often tried to define to myself the scent of the Calycanthus blooms; they have an aromatic fragrance somewhat like the ripest Pineapples of the tropics, but still richer; how I love to carry them in my hand, crushed and warm, occasionally holding them tight over my mouth and nose to fill myself with their perfume. The leaves have a similar, but somewhat varied and sharper, scent, and the woody stems another; the latter I like to nibble. This flower has an element of mystery in it — that indescribable quality felt by children, and remembered by prosaic grown folk. Perhaps its curious dark reddish brown tint may have added part of the queerness, since the "Mourning Bride," similar in color, has a like mysterious association. I cannot explain these qualities to any one not a garden-bred child; and as given in the chapter entitled The Mystery of Flowers, they will appear to many, fanciful and unreal — but I have a fraternity who will understand,

and who will know that it was this same undefinable quality that made a branch of Strawberry bush, or a handful of its stemless blooms, a gift significant of interest and intimacy; we would not willingly give

Hawthorn Arch at Holly House, Peace Dale, Rhode Island.
Home of Rowland G. Hazard, Esq.

Calycanthus blossoms to a child we did not like, or to a stranger.

A rare perfume floats from the modest yellow Flowering Currant. I do not see this sweet and sightly shrub in many modern gardens, and it is our loss. The crowding bees are goodly and cheerful, and the flowers are pleasant, but the perfume is of the sort you can truly say you love it; its aroma is like some of the liqueurs of the old monks.

The greatest pleasure in flower perfumes comes to us through the first flowers of spring. How we breathe in their sweetness! Our native wild flowers give us the most delicate odors. The May-flower is, I believe, the only wild flower for which all country folk of New England have a sincere affection; it is not only a beautiful, an enchanting flower, but it is so fresh, so balmy of bloom. It has the delicacy of texture and form characteristic of many of our native spring blooms, Hepatica, Anemone, Spring Beauty, Polygala.

The Arethusa was one of the special favorites of my father and mother, who delighted in its exquisite fragrance. Hawthorne said of it: "One of the delicatest, gracefullest, and in every manner sweetest of the whole race of flowers. For a fortnight past I have found it in the swampy meadows, growing up to its chin in heaps of wet moss. Its hue is a delicate pink, of various depths of shade, and somewhat in the form of a Grecian helmet."

It pleases me to fancy that Hawthorne was like the Arethusa, that it was a fit symbol of the nature of our greatest New England genius. Perfect in grace and beauty, full of sentiment, classic and elegant of shape, it has a shrinking heart; the sepals and petals rise over it and shield it, and the whole flower is shy and retiring, hiding in marshes and quaking bogs.

It is one of our flowers which we ever regard singly, as an individual, a rare and fine spirit; we never think of it as growing in an expanse or even in groups. This lovely flower has, as Landor said of the flower of the vine, "a scent so delicate that it requires a sigh to inhale it."

The faintest flower scents are the best. You find yourself longing for just a little more, and you bury your face in the flowers and try to draw out a stronger breath of balm. Apple blossoms, certain Violets, and Pansies have this pale perfume.

In the front yard of my childhood's home grew a Larch, an exquisitely graceful tree, one now little planted in Northern climates. I recall with special delight the faint fragrance of its early shoots. The next tree was a splendid pink Hawthorn. What a day of mourning it was when it had to be cut down, for trees had been planted so closely that many must be sacrificed as years went on and all grew in stature.

There are some smells that are strangely pleasing to the country lover which are neither from fragrant flower nor leaf; one is the scent of the upturned earth, most heartily appreciated in early spring. The smell of a ploughed field is perhaps the best of all earthy scents, though what Bliss Carman calls "the racy smell of the forest loam" is always good. Another is the burning of weeds of garden rakings,

"The spicy smoke
Of withered weeds that burn where gardens be."

A garden "weed-smother" always makes me think of my home garden, and my father, who used to stand by this burning weed-heap, raking in the withered leaves. Many such scents are pleasing chiefly through the power of association.

Thyme-covered Graves.

The sense of smell in its psychological relations is most subtle: —

"The subtle power in perfume found,
Nor priest nor sibyl vainly learned;
On Grecian shrine or Aztec mound
No censer idly burned.

"And Nature holds in wood and field
Her thousand sunlit censers still;
To spells of flower and shrub we yield
Against or with our will."

Dr. Holmes notes that memory, imagination, sentiment, are most readily touched through the sense of smell. He tells of the associations borne to him by the scent of Marigold, of Life-everlasting, of an herb closet.

Notwithstanding all these tributes to sweet scents and to the sense of smell, it is not deemed, save in poetry, wholly meet to dwell much on smells, even pleasant ones. To all who here sniff a little disdainfully at a whole chapter given to flower scents, let me repeat the Oriental proverb: —

> "To raise Flowers is a Common Thing,
> God alone gives them Fragrance."

Balmier far, and more stimulating and satisfying than the perfumes of most blossoms, is the scent of aromatic or balsamic leaves, of herbs, of green growing things. Sweetbrier, says Thoreau, is thus "thrice crowned: in fragrant leaf, tinted flower, and glossy fruit." Every spring we long, as Whittier wrote —

> "To come to Bayberry scented slopes,
> And fragrant Fern and Groundmat vine,
> Breathe airs blown o'er holt and copse,
> Sweet with black Birch and Pine."

All these scents of holt and copse are dear to New Englanders.

I have tried to explain the reason for the charm to me of growing Thyme. It is not its beautiful perfume, its clear vivid green, its tiny fresh flowers, or the element of historic interest. Alphonse Karr gives another reason, a sentiment of gratitude. He says: —

"Thyme takes upon itself to embellish the parts of the earth which other plants disdain. If there is an arid, stony, dry soil, burnt up by the sun, it is there Thyme spreads its charming green beds, perfumed, close, thick, elastic, scattered over with little balls of blossom, pink in color, and of a delightful freshness."

Thyme was, in older days, spelt Thime and Time. This made the poet call it "pun-provoking Thyme." I have an ancient recipe from an old herbal for "Water of Time to ease the Passions of the Heart." This remedy is efficacious to-day, whether you spell it time or thyme.

There are shown on page 301 some lonely graves in the old Moravian burying-ground in Bethlehem, overgrown with the pleasant perfumed Thyme. And as we stand by their side we think with a half smile — a tender one — of the never-failing pun of the old herbalists.

Spenser called Thyme "bee-alluring," "honey-laden." It was the symbol of sweetness; and the Thyme that grew on the sunny slopes of Mt. Hymettus gave to the bees the sweetest and most famed of all honey. The plant furnished physic as well as perfume and puns and honey. Pliny named eighteen sovereign remedies made from Thyme. These cured everything from the "bite of poysonful spidars" to "the Apoplex." There were so many recipes in the English *Compleat Chirurgeon*, and similar medical books, that you would fancy venomous spiders were as thick as gnats in England. These spider cure-alls are however simply a proof that the recipes were taken from dose-books of Pliny and various Roman physicians, with whom spider bites were more common and more painful than in England.

The Haven of Health, written in 1366, with a special view to the curing of "Students," says that Wild Thyme has a great power to drive away heaviness of mind, "to purge melancholly and splenetick humours." And the author recommends to "sup the leaves with eggs." The leaves were used everywhere "to be put in puddings and such like meates, so that in divers places Thime was called Pudding-grass." Pudding in early days was the stuffing of meat and poultry, while concoctions of eggs, milk, flour, sugar, etc., like our modern puddings, were called whitpot.

Many traditions hang around Thyme. It was used widely in incantations and charms. It was even one of the herbs through whose magic power you could see fairies. Here is a "Choice Proven Secret made Known" from the Ashmolean Mss.

How to see Fayries

"℞. A pint of Sallet-Oyle and put it into a vial-glasse but first wash it with Rose-water and Marygolde-water the Flowers to be gathered toward the East. Wash it until teh Oyle come white. Then put it in the glasse, *ut supra:* Then put thereto the budds of Holyhocke, the flowers of Marygolde, the flowers or toppers of Wild Thyme, the budds of young Hazle: and the time must be gathered neare the side of a Hill where Fayries used to be: and take the grasse off a Fayrie throne. Then all these put into the Oyle into the Glasse, and sette it to dissolve three dayes in the Sunne and then keep for thy use *ut supra*."

"I know a bank whereon the Wild Thyme blows"—it is not in old England, but on Long

Island ; the dense clusters of tiny aromatic flowers form a thick cushioned carpet under our feet. Lord Bacon says in his essay on Gardens : —

"Those which perfume the air most delightfully, not passed by as the rest, but being trodden upon and crushed

"White Umbrellas of Elder."

are three : that is, Burnet, Wild Thyme, and Water-Mints. Therefore you are to set whole alleys of them, to have the pleasure when you walk or tread."

Here we have an alley of Thyme, set by nature, for us to tread upon and enjoy, though Thyme always seems to me so classic a plant, that it is far too fine to walk upon ; one ought rather to sleep and dream upon it.

Great bushes of Elder, another flower of witchcraft, grow and blossom near my Thyme bank. Old Thomas Browne, as long ago as 1685 called the Elder bloom "white umbrellas" — which has puzzled me much, since we are told to assign the use and knowledge of umbrellas in England to a much later date; perhaps he really wrote umbellas. Now it is a well-known fact — sworn to in scores of old herbals, that any one who stands on Wild Thyme, by the side of an Elder bush, on Midsummer Eve, will "see great experiences"; his eyes will be opened, his wits quickened, his vision clarified; and some have even seen fairies, pixies — Shakespeare's elves — sporting over the Thyme at their feet.

I shall not tell whom I saw walking on my Wild Thyme bank last Midsummer Eve. I did not need the Elder bush to open my eyes. I watched the twain strolling back and forth in the half-light, and I heard snatches of talk as they walked toward me, and I lost the responses as they turned from me. At last, in a louder voice: —

HE. "What is this jolly smell all around here? Just like a mint-julep! Some kind of a flower?"

SHE. "It's Thyme, Wild Thyme; it has run into the edge of the lawn from the field, and is just ruining the grass."

HE (*stooping to pick it*). "Why, so it is. I thought it came from that big white flower over there by the hedge."

SHE. "No, that is Elder."

HE (*after a pause*). "I had to learn a lot of old Arnold's poetry at school once, or in college, and there was some just like to-night: —

"'The evening comes — the fields are still,
The tinkle of the thirsty rill,
Unheard all day, ascends again.
Deserted is the half-mown plain,
And from the Thyme upon the height,
And from the Elder-blossom white,
And pale Dog Roses in the hedge,
And from the Mint-plant in the sedge,
In puffs of balm the night air blows
The perfume which the day foregoes —
And on the pure horizon far
See pulsing with the first-born star
The liquid light above the hill.
The evening comes — the fields are still.'"

Then came the silence and half-stiffness which is ever apt to follow any long quotation, especially any rare recitation of verse by those who are notoriously indifferent to the charms of rhyme and rhythm, and are of another sex than the listener. It seems to indicate an unusual condition of emotion, to be a sort of barometer of sentiment, and the warning of threatening weather was not unheeded by her; hence her response was somewhat nervous in utterance, and instinctively perverse and contradictory.

SHE. "That line, 'The liquid light above the hill,' is very lovely, but I can't see that it's any of it at all like to-night."

HE (*stoutly and resentfully*). "Oh, no! not at all! There's the field, all still, and here's Thyme, and Elder, and there are wild Roses! — and see! the moon is coming up — so there's your liquid light."

SHE. "Well! Yes, perhaps it is; at any rate it is a lovely night. You've read *Lavengro?* No? Certainly you must have heard of it. The gipsy in it says: 'Life is sweet, brother. There's day and night, brother, both sweet things; sun, moon, and stars, brother, all sweet things; there is likewise a wind on the heath.'"

He (*dubiously*). "That's rather queer poetry, if it is poetry —and you must know I do not like to hear you call me brother."

Whereupon I discreetly betrayed my near presence on the piazza, to prove that the field, though still, was not deserted. And soon the twain said they would walk to the club house to view the golf prizes; and they left the Wild Thyme and Elder blossoms white, and turned their backs on the moon, and fell to golf and other eminently unromantic topics, far safer for Midsummer Eve than poesy and other sweet things.

Lower Garden at Sylvester Manor.

CHAPTER XIV

JOAN SILVER-PIN

"Being of many variable colours, and of great beautie, although of evill smell, our gentlewomen doe call them Jone Silver-pin."

—John Gerarde, *Herball*, 1596.

ARDEN Poppies were the Joan Silver-pin of Gerarde, stigmatized also by Parkinson as "Jone Silver-pinne, *subauditur*; faire without and foule within." In Elizabeth's day Poppies met universal distrust and aversion, as being the source of the dreaded opium. Spenser called the flower "dead-sleeping" Poppy; Morris "the black heart, amorous Poppy"—which might refer to the black spots in the flower's heart.

Clare, in his *Shepherd's Calendar* also asperses them:—

"Corn-poppies, that in crimson dwell,
Called Head-aches from their sickly smell."

Forby adds this testimony: "Any one by smelling of it for a very short time may convince himself of the propriety of the name." Some fancied that the dazzle of color caused headaches—that vivid scarlet, so fine a word as well as color that it is annoying to hear the poets change it to crimson.

This regard of and aversion to the Poppy lingered among elderly folks till our own day; and I well recall the horror of a visitor of antique years in our mother's garden during our childhood, when we were found cheerfully eating Poppy seeds. She viewed us with openly expressed apprehension that

"Black Heart, Amorous Poppies."

we would fall into a stupor; and quite terrified us and our relatives, in spite of our assertions that we "always ate them," which indeed we always did and do to this day; and very pleasant of taste they are, and of absolutely no effect, and not at all of evil smell to our present fancy, either in blossom or seed, though distinctly medicinal in odor.

Returned missionaries were frequent and honored visitors in our town and our house in those days; and one of these good men reassured us and reinstated in favor our uncanny feast by telling us that in the East, Poppy seeds were eaten everywhere, and were frequently baked with wheaten flour into cakes. A dislike of the scent of Field Poppies is often found among English folk. The author of *A World in a Garden* speaks in disgust of "the pungent and sickly odor of the flaring Poppies — they positively nauseate me"; but then he disliked their color too.

There is something very fine about a Poppy, in the extraordinary combination of boldness of color and great size with its slender delicacy of stem, the grace of the set of the beautiful buds, the fine turn of the flower as it opens, and the wonderful airiness of poise of so heavy a flower. The silkiness of tissue of the petals, and their semi-transparency in some colors, and the delicate fringes of some varieties, are great charms.

> Each crumpled crêpe-like leaf is soft as silk;
> Long, long ago the children saw them there,
> Scarlet and rose, with fringes white as milk,
> And called them 'shawls for fairies' dainty wear';
> They were not finer, those laid safe away
> In that low attic, neath the brown, warm eaves."

And when the flowers have shed, oh, so lightly! their silken petals, there is still another beauty, a seed vessel of such classic shape that it wears a crown.

I have always rejoiced in the tributes paid to the Poppy by Ruskin and Mrs. Thaxter. She deemed them the most satisfactory flower among the annuals "for wondrous variety, certain picturesque qualities, for color and form, and a subtle air of mystery."

There is a line of Poppy colors which is most entrancing; the gray, smoke color, lavender, mauve, and lilac Poppies, edged often and freaked with tints of red, are rarely beautiful things. There are fine white Poppies, some fringed, some single, some double — the Bride is the appropriate name of the fairest. And the pinks of Poppies, that wonderful red-pink, and a shell-pink that is almost salmon, and the sunset pinks of our modern Shirley Poppies, with quality like finest silken gauze! The story of the Shirley Poppies is one of magic, that a flower-loving clergyman who in 1882 sowed the seed of one specially beautiful Poppy which had no black in it, and then sowed those of its fine successors, produced thus a variety which has supplied the world with beauty. Rev. Mr. Wilks, their raiser, gives these simply worded rules anent his Shirley Poppies: —

> "1, They are single; 2, always have a white base; 3, with yellow or white stamens, anthers, or pollen; 4, and never have the smallest particle of black about them."

The thought of these successful and beautiful Poppies is very stimulating to flower raisers of moderate means, with no profound knowledge of flowers; it shows what can be done by enthusiasm and application and patience. It gives something of the same comfort found in Keats's fine lines to the singing thrush: —

> "Oh! fret not after knowledge.
> I have none, *and yet the evening listens.*"

Notwithstanding all this distinction and beauty, these fine things of the garden were dubbed Joan Silver-pin. I wonder who Joan Silver-pin was! I have searched faithfully for her, but have not been able to get on the right scent. Was she of real life, or fiction? I have looked through the lists of characters of contemporary plays, and read a few old jest books and some short tales of that desperately colorless sort, wherein you read page after page of the printed words with as little absorption of signification as if they were Choctaw. But never have I seen Joan Silver-pin's name; it was a bit of Elizabethan slang, I suspect, — a cant term once well known by every one, now existing solely through this chance reference of the old herbalists.

No garden can aspire to be named An Old-fashioned Garden unless it contains that beautiful plant the Garden Valerian, known throughout New England to-day as Garden Heliotrope; as Setwall it grew in every old garden, as it was in every pharmacopœia. It was termed "drink-quickening Setuale" by Spenser, from the universal use of its flowers to flavor various enticing drinks. Its lovely blossoms are pinkish in bud and open to pure white; its curiously penetrating vanilla-like fragrance is disliked by many who are not cats. I find it rather pleasing of scent when growing in the garden, and not at all like the extremely nasty-smelling medicine which is made from it, and which has been used for centuries for "histerrick fits," and is still constantly prescribed to-day for that unsympathized-with malady. Dr. Holmes calls it, "Valerian, calmer of hysteric

Valerian.

squirms." It is a stately plant when in tall flower in June; my sister had great clumps of bloom like the ones shown above, but alas! the cats caught them before the photographer did. The cats did not have to watch the wind and sun and rain, to pick out plates and pack plate-holders, and gather ray-fillers and cloth and lens, and adjust the tripod, and fix the camera and focus, and think, and focus, and think, and then wait — till the wind ceased blowing. So when they found it, they broke down every slender stalk and rolled in it till the ground was tamped down as hard as if one of our lazy road-menders had been at it. Valerian has in England as an appropriate folk name, "Cats'-fancy." The pretty little annual, Nemophila, makes also a favorite rolling-place for our cat; while all who love cats have given them Catnip and seen the singular intoxication it brings. The sight of a cat in this strange ecstasy over a bunch of Catnip always gives me a half-sense of fear; she becomes such a truly wild creature, such a miniature tiger.

In *The Art of Gardening*, by J. W., Gent., 1683, the author says of Marigolds: "There are divers sorts besides the common as the African Marigold, a Fair bigge Yellow Flower, but of a very Naughty Smell." I cannot refrain, ere I tell more of the Marigold's naughtiness, to copy a note written in this book by a Massachusetts bride whose new husband owned and studied the book two hundred years ago; for it gives a little glimpse of old-time life. In her exact little handwriting are these words: —

"Planted in Potts, 1720: An Almond Stone, an English Wallnut, Cittron Seeds, Pistachica nutts, Red Damsons, Leamon seeds, Oring seeds and Daits."

Poor Anne! she died before she had time to become any one's grandmother. I hope her successor in matrimony, our forbear, cherished her little seedlings and rejoiced in the Lemon and Almond trees, though Anne herself was so speedily forgotten. She is, however, avenged by Time; for she is remembered better than the wife who took her place, through her simple flower-loving words.

I am surprised at this aspersion on the Marigold as to its smell, for all the traditions of this flower show it to have been a great favorite in kitchen gardens; and I have found that elderly folk are very apt to like its scent. My father loved the flower and the fragrance, and liked to have a bowl of Marigolds stand beside him on his library table. It was constantly carried to church as a "Sabbath-day posy," and its petals used as flavoring in soups and stews. Charles Lamb said it poisoned them. Canon Ellacombe writes that it has been banished in England to the gardens of cottages and old farm-houses; it had a waning popularity in America, but was never wholly despised.

How Edward Fitzgerald loved the African Marigold! "Its grand color is so comfortable to us Spanish-like Paddies," he writes to Fanny Kemble in letters punctuated with little references to his garden flowers: letters so cheerful, too, with capitals; "I love the old way of Capitals for Names," he says — and so do I; letters bearing two surprises, namely, the infrequent references to Omar Khayyam; and the fact that Nasturtiums, not Roses, were his favorite flower.

The question of the agreeableness of a flower scent is a matter of public opinion as well as personal choice. Environment and education influence us. In olden times every one liked certain scents deemed odious to-day. Parkinson's praise of Sweet Sultans was, "They are of so exceeding sweet a scent as it surpasses the best civet that is." Have you ever smelt civet? You will need no words to tell you that the civet is a little cousin of the skunk. Cowper could not talk with civet in the room; most of us could not even breathe. The old herbalists call Privet sweet-scented. I don't know that it is strange to find a generation who loved civet and musk thinking Privet pleasant-scented. Nearly all our modern botanists have copied the words of their predecessors; but I scarcely know what to say or to think when I find so exact an observer as John Burroughs calling Privet "faintly sweet-scented." I find it rankly ill-scented.

The men of Elizabethan days were much more learned in perfumes and fonder of them than are most folk to-day. Authors and poets dwelt frankly upon them without seeming at all vulgar. Of course herbalists, from their choice of subject, were free to write of them at length, and they did so with evident delight. Nowadays the French realists are the only writers who boldly reckon with the sense of smell. It isn't deemed exactly respectable to dwell too much on smells, even pleasant ones; so this chapter certainly must be brief.

I suppose nine-tenths of all who love flower scents would give Violets as their favorite fragrance;

yet how quickly, in the hothouse Violets, can the scent become nauseous. I recall one formal luncheon whereat the many tables were mightily massed with violets; and though all looked as fresh as daybreak to the sight, some must have been gathered for a day or more, and the stale odor throughout the room was unbearable. But it is scarcely fair to decry a flower because of its scent in decay. Shakespeare wrote: —

> "Lilies festered smell far worse than weeds."

Many of our Compositæ are vile after standing in water in vases; Ox-eye Daisies, Rudbeckia, Zinnia, Sunflower, and even the wholesome Marigold. Delicate as is the scent of the Pansy, the smell of a bed of ancient Pansy plants is bad beyond words. The scent of the flowers of fruit-bearing trees is usually delightful; but I cannot like the scent of Pear blossoms.

I dislike much the rank smell of common yellow Daffodils and of many of that family. I can scarcely tolerate them even when freshly picked, upon a dinner table. Some of the Jonquils are as sickening within doors as the Tuberose, though in both cases it is only because the scent is confined that it is cloying. In the open air, at a slight distance, they smell as well as many Lilies, and the Poet's Narcissus is deemed by many delightful.

I have ever found the scent of Lilacs somewhat imperfect, not well rounded, not wholly satisfying; but one of my friends can never find in a bunch of our spring Lilacs any odor save that of illuminating gas. I do wish he had not told me this! Now when I stand beside my Lilac bush I feel like looking around anxiously to see where the gas is escaping. Linnæus thought the perfume of Mignonette the purest ambrosia. Another thinks that Mignonette has a doggy smell, as have several flowers; this is not wholly to their disparagement. Our cocker spaniel is sweeter than some flowers, but he is not a Mignonette. There be those who love most of all the scent of Heliotrope, which is to me a close, almost musty scent. I have even known of one or two who disliked the scent of Roses, and the Rose itself has been abhorred. Marie de' Medici would not even look at a painting or carving of a Rose. The Chevalier de Guise, had a loathing for Roses. Lady Heneage, one

Old "War Office."

of the maids of honor to Queen Elizabeth, was made very ill by the presence or scent of Roses. This illness was not akin to "Rose cold," which is the baneful companion of so many Americans, and which can conquer its victims in the most sudden and complete manner.

Even my affection for Roses, and my intense love of their fragrance, shown in its most ineffable sweetness in the old pink Cabbage Rose, will not cause me to be silent as to the scent of some of the Rose sisters. Some of the Tea Roses, so lovely of texture, so delicate of hue, are sickening; one has a suggestion of ether which is most offensive. "A Rose by any other name would smell as sweet," but not if its name (and its being) was the Persian Yellow. This beautiful double Rose of rich yellow was introduced to our gardens about 1830. It is infrequent now, though I find it in florists' lists; and I suspect I know why. Of late years I have not seen it, but I have a remembrance of its uprootal from our garden. Mrs. Wright confirms my memory by calling it "a horrible thing — the Skunk Cabbage of the garden." It smells as if foul insects were hidden within it, a disgusting smell. I wonder whether poor Marie de' Medici hadn't had a whiff of it. A Persian Rose! it cannot be possible that Omar Khayyam ever smelt it, or any of the Rose singers of Persia, else their praises would have turned to loathing as they fled from its presence. There are two or three yellow Roses which are not pleasing, but are not abhorrent as is the Persian Yellow.

One evening last May I walked down the garden

path, then by the shadowy fence-side toward the barn. I was not wandering in the garden for sweet moonlight, for there was none; nor for love of flowers, nor in admiration of any of nature's works, for it was very cold; we even spoke of frost, as we ever do apprehensively on a chilly night in spring. The kitten was lost. She was in the shrubbery at the garden end, for I could hear her plaintive yowling; and I thus traced her. I gathered her up, purring and clawing, when I heard by my side a cross rustling of leaves and another complaining voice. It was the Crown-imperial, unmindful or unwitting of my presence, and muttering peevishly: "Here I am, out of fashion, and therefore out of the world! torn away from the honored border by the front door path, and even set away from the broad garden beds, and thrust with sunflowers and other plants of no social position whatever down here behind the barn, where, she dares to say, we 'can all smell to heaven together.'

"What airs, forsooth! these twentieth century children put on! Smell to heaven, indeed! I wish her grandfather could have heard her! He didn't make such a fuss about smells when I was young, nor did any one else; no one's nose was so over-nice. Every spring when I came up, glorious in my dress of scarlet and green, and hung with my jewels of pearls, they were all glad to see me and to smell me, too; and well they might be, for there was a rotten-appley, old-potatoey smell in the cellar which pervaded the whole house when doors were closed. And when the frost came up from the ground the old sink drain at the kitchen door rendered up to the spring sunshine all the combined vapors of all the dish-water of all the winter. The barn and hen-house and cow-house reeked in the sunlight, but the pigpen easily conquered them all. There was an ancient cesspool far too near the kitchen door, underground and not to be seen, but present, nevertheless. A hogshead of rain-water stood at the cellar door, and one at the end of the barn — to water the flowers with — they fancied rotten rain-water made flowers grow! A foul dye-tub was ever reeking in every kitchen chimney corner, a culminating horror in stenches; and vessels of ancient soap grease festered in the outer shed, the grease collected through the winter and waiting for the spring soap-making. The vapor of sour milk, ever present, was of little moment — when there was so much else so much worse. There wasn't a bath-tub in the grandfather's house, nor in any other house in town, nor any too much bathing in winter, either, I am sure, in icy well-water in icier sleeping rooms. The windows were carefully closed all winter long, but the open fireplaces managed to save the life of the inmates, though the walls and rafters were hung with millions of germs which every one knows are all the wickeder when they don't smell, because you take no care, fancying they are not there. But the grandfather knew naught of germs — and was happy. The trees shaded the house so that the roof was always damp. Oh, how those germs grew and multiplied in the grateful shade of those lovely trees, and how mould and rust rejoiced. Well might people turn from all these sights and scents to me. The grandfather and his wife, when they were young, as when they were in middle age, and when they were old, walked every early spring day at set of sun, slowly down the front path, looking at every flower, every bud; pulling a tiny weed, gathering a choice flower, breaking a withered sprig; and they ever lingered long and happily by my side. And he always said, 'Wife! isn't this Crown-imperial a glorious plant? so stately, so perfect in form, such an expression of life, and such a personification of spring!' 'Yes, father,' she would answer quickly, 'but don't pick it.' Why, I should have resented even that word had she referred to my perfume. She meant that the garden border could not spare me. The children never could pick me, even the naughtiest ones did not dare to; but they could pull all the little upstart Ladies' Delights and Violets they wished. And yet, with all this family homage which should make me a family totem, here I am, stuck down by the barn—I, who sprung from the blood of a king, the great Gustavus Adolphus—and was sung by a poet two centuries ago in the famous *Garland of Julia*. The old Jesuit poet Rapin said of me, 'No flower aspires in pomp and state so high.'

"Read this page from that master-herbalist, John Gerarde, telling of the rare beauties within my golden cup.

"A very intelligent and respectable old gentleman named Parkinson, who knew far more about flowers than flighty folk do nowadays, loved me well and wrote of me, 'The Crown-imperial, for its stately beautifulnesse deserveth the first place in this our garden of delight to be here entreated of before all other Lilies.' He had good sense. It was not I who was stigmatized by him as Joan Silver-pin. He spoke very plainly and very sensibly of my perfume; there was no nonsense in his notions, he told the truth, the whole truth, and nothing but the truth: 'The whole plant and every part thereof, as well as rootes as leaves and floures doe smell somewhat strong, as it were the savour of a foxe, so that if any doe but near it, he can but smell it, yet is not unwholesome.'

"How different all is to-day in literature, as well as in flower culture. Now there are low, coarse attempts at wit that fairly wilt a sensitive nature like mine. There is one miserable Man who comes to this garden, and who *thinks* he is a Poet; I will not repeat his wretched rhymes. But only yesterday, when he stood looking superciliously down upon us, he said sneeringly, 'Yes, spring is here, balmy spring; we know her presence without seeing her face or hearing her voice; for the Skunk Cabbage is unfurled in the swamps, and the Crown-imperial is blooming in the garden.' Think of his presuming to set me alongside that low Skunk Cabbage — me with my 'stately beautifulness.'

"Little do people nowadays know about scents anyway, when their botanists and naturalists write that the Privet bloom is 'pleasingly fragrant,' and one dame set last summer a dish of Privet on her dining table before many guests. Privet! with its ancient and fishlike smell! And another tells

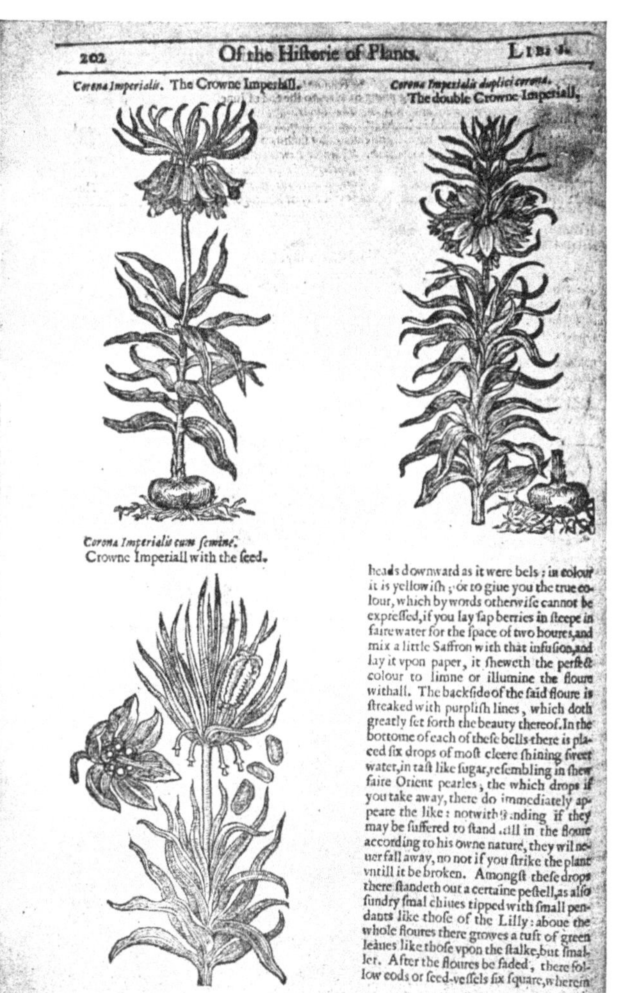

202 Of the Historie of Plants. Lib. 1.

Corona Imperialis. The Crowne Imperiall.

Corona Imperialis duplici corona. The double Crowne Imperiall.

Corona Imperialis cum semine. Crowne Imperiall with the seed.

heads downward as it were bels: in colour it is yellowish; or to giue you the true colour, which by words otherwise cannot be expressed, if you lay sap berries in steepe in faire water for the space of two houres, and mix a little Saffron with that infusion, and lay it vpon paper, it sheweth the perfect colour to limne or illumine the floure withall. The backside of the said floure is streaked with purplish lines, which doth greatly set forth the beauty thereof. In the bottome of each of these bells there is placed six drops of most cleere shining sweet water, in tast like sugar, resembling in shew faire Orient pearles, the which drops if you take away, there do immediately appeare the like: notwithstanding if they may be suffered to stand still in the floure according to his owne nature, they wil neuer fall away, no not if you strike the plant vntill it be broken. Amongst these drops there standeth out a certaine pestell, as also sundry smal chiues tipped with small pendants like those of the Lilly: aboue the whole floures there growes a tuft of green leaues like those vpon the stalke, but smaller. After the floures be faded, there follow cods or seed-vessels six square, wherein is

Crown Imperial. A Page from Gerarde's *Herball.*

of the fragrant delight of flowering Buckwheat—may the breezes blow such fragrance far from me! But why dwell on perfumes; flowers were made to look at, not to smell; sprays of Sweet Balm or Basil leaves outsweeten every flower, and make no pretence or thought of beauty; render to each its own virtues, and try not to engross the charm of another.

"I was indeed the queen of the garden, and here I am exiled behind the barn. Life is not worth living. I won't come up again. She will walk through the garden next May and say, 'How dull and shabby the garden looks this year! the spring is backward, everything has run to leaves, nothing is in bloom, we must buy more fertilizer, we must get a new gardener, we must get more plants and slips and seeds and bulbs, it is fearfully discouraging, I never saw anything so gone off!' then perhaps she will remember, and regret the friend of her grandparents, the Crown-imperial—whom she thrust from her Garden of Delight."

CHAPTER XV

CHILDHOOD IN A GARDEN

"I see the garden thicket's shade
Where all the summer long we played,
And gardens set and houses made,
Our early work and late."

— MARY HOWITT.

OW we thank God for the noble traits of our ancestors; and our hearts fill with gratitude for the tenderness, the patience, the loving kindness of our parents; I have an infinite deal for which to be sincerely grateful; but for nothing am I now more happy than that there were given to me a flower-loving father and mother. To that flower-loving father and mother I offer in tenderest memory equal gratitude for a childhood spent in a garden.

Winter as well as summer gave us many happy garden hours. Sometimes a sudden thaw of heavy snow and an equally quick frost formed a miniature pond for sheltered skating at the lower end of the garden. A frozen crust of snow (which our winters nowadays so seldom afford) gave other joys. And the delights of making a snow man, or a snow fort, even of rolling great globes of snow, were infinite and varied. More subtle was the charm of shaping certain *things* from dried twigs and evergreen sprigs, and pouring water over them to freeze into a beautiful resemblance of the original form. These might be the ornate initials or name of a dear girl friend, or a tiny tower or pagoda. I once had a real winter garden in miniature set in twigs of cedar and spruce, and frozen into a fairy garden.

In summertime the old-fashioned garden was a paradise for a child; the long warm days saw the fresh telling of child to child, by that curiously subtle system of transmission which exists everywhere among happy children, of quaint flower customs known to centuries of English-speaking children, and also some newer customs developed by the fitness of local flowers for such games and plays.

The Countess Potocka says the intense enjoyment of nature is a sixth sense. We are not born with this good gift, nor do we often acquire it in later life; it comes through our rearing. The fulness of delight in a garden is the bequest of a childhood spent in a garden. No study or possession of flowers in mature years can afford gratification equal to that conferred by childish associations with them; by the sudden recollection of flower lore, the memory of child friendships, the recalling of games or toys made of flowers: you cannot explain it; it seems a concentration, an extract of all the sunshine and all the beauty of those happy summers of our lives when the whole day and every day was spent among flowers. The sober

Milkweed Seed.

teachings of science in later years can never make up the loss to children debarred of this inheritance, who have grown up knowing not when "the summer comes with bee and flower."

A garden childhood gives more sources of delight to the senses in after life than come from beautiful color and fine fragrance. Have you pleasure in the contact of a flower? Do you like its touch as well as its perfume? Do you love to feel a Lilac spray brush your cheek in the cool of the evening? Do you like to bury your face in a bunch of Roses? How frail and papery is the Larkspur! And how silky is the Poppy! A Locust bloom is a fringe of sweetness; and how very doubtful is the touch of the Lily — an unpleasant thick sleekness. The Clove Carnation is the best of all. It feels just as it smells. These and scores more give me pleasure through their touch, the result of constant handling of flowers when I was a child.

There were harmful flowers in the old garden — among them the Monk's-hood; we never touched it, except warily. Doubtless we were warned, but we knew it by instinct and did not need to be told. I always used to see in modest homes great tubs each with a flourishing Oleander tree. I have set out scores of little slips of Oleander, just as I planted Orange seeds. I seldom see Oleanders now; I wonder whether the plant has been banished on account of its poisonous properties. I heard of but one fatal case of Oleander poisoning — and that was doubtful. A little child, the sister of one of my playmates, died suddenly in great distress. Several months after her death the mother was told that the leaves of the Oleander were poisonous, when she recalled that the child had eaten them on the day of her death.

Oleander blossoms were lovely in shape and color. Edward Fitzgerald writes to Fanny Kemble: "Don't you love the Oleander? So clean in its Leaves and Stem, as so beautiful in its Flower; loving to stand in water which it drinks up fast. I have written all my best Mss. with a Pen that has been held with its nib in water for more than a fortnight — Charles Keene's recipe for keeping Pens in condition — Oleander-like." This, written in 1882, must, even at that recent date, refer to quill pens.

The lines of Mary Howitt's, quoted at the beginning of this chapter, ring to me so true; there is in them no mock sentiment, it is the real thing, — "the garden thicket's shade," little "cubby houses" under the close-growing stems of Lilac and Syringa, with an old thick shawl outspread on the damp earth for a carpet. Oh, how hot and scant the air was in the green light of those close "garden-thickets," those "Lilac ambushes," which were really not half so pleasant as the cooler seats on the grass under the trees, but which we clung to with a warmth equal to their temperature.

Let us peer into these garden thickets at these happy little girls, fantastic in their garden dress. Their hair is hung thick with Dandelion curls, made from pale green opal-tinted stems that have grown long under the shrubbery and Box borders. Around their necks are childish wampum, strings of Dandelion beads or Daisy chains. More delicate wreaths for the neck or hair were made from the

The Children's Garden.

blossoms of the Four-o'clock or the petals of Phlox or Lilacs, threaded with pretty alternation of color. Fuchsias were hung at the ears for eardrops, green leaves were pinned with leaf stems into little caps and bonnets and aprons, Foxgloves made dainty children's gloves. Truly the garden-bred child went in gay attire.

That exquisite thing, the seed of Milkweed (shown on page 328), furnished abundant playthings. The plant was sternly exterminated in our garden, but sallies into a neighboring field provided supplies for fairy cradles with tiny pillows of silvery silk.

One of the early impulses of infancy is to put everything in the mouth; this impulse makes the creeping days of some children a period of constant watchfulness and terror to their apprehensive guardians. When the children are older and can walk in the garden or edge of the woods, a fresh anxiety arises; for a certain savagery in their make-up makes them regard every growing thing, not as an object to look at or even to play with, but to eat. It is a relief to the mother when the child grows beyond the savage, and falls under the dominion of tradition and folklore, communicated to him by other children by that subtle power of enlightenment common to children, which seems more like instinct than instruction. The child still eats, but he makes distinctions, and seldom touches harmful leaves or seeds or berries. He has an astonishing range: roots, twigs, leaves, bark, tendrils, fruit, berries, flowers, buds, seeds, all alike serve for food. Young shoots of Sweetbrier and Blackberry are nibbled as well as the branches of young Birch. Grape tendrils, too, have an acid zest, as do Sorrel leaves. Wild Rose hips and the drupes of dwarf Cornel are chewed. The leaf buds of Spruce and Linden are also tasted. I hear that some children in some places eat the young fronds of Cinnamon Fern, but I never saw it done. Seeds of Pumpkins and Sunflowers were edible, as well as Hollyhock cheeses. There was one Slippery Elm tree which we know in our town, and we took ample toll of it. Cherry gum and Plum gum are chewed, as well as the gum of Spruce trees. There was a boy who used sometimes to intrude on our girl's paradise, since he was the son of a neighbor, and he said he ate raw Turnips, and something he called Pig-nuts — I wonder what they were.

Those childish customs linger long in our minds, or rather in our subconsciousness. I never walk through an old garden without wishing to nibble and browse on the leaves and stems which I ate as a child, without sucking a drop of honey from certain flowers. I do it not with intent, but I waken to realization with the petal of Trumpet Honeysuckle in my hand and its drop of ambrosia on my lips.

Children care far less for scent and perfection in a flower than they do for color, and, above all, for desirability and adaptability of form, this desirability being afforded by the fitness of the flower for the traditional games and plays. The favorite flowers of my childhood were three noble creatures, Hollyhocks, Canterbury Bells, and Foxgloves, all three were scentless. I cannot think of a child's summer in a garden without these three old favorites of history and folk-lore. Of course we enjoyed the earlier flower blooms and played happily with them ere our dearest treasures came to us; but never had we full variety, zest, and satisfaction till this trio were

Foxgloves in a Narragansett Garden.

in midsummer bloom. There was a little gawky, crudely-shaped wooden doll of German manufacture sold in Worcester which I never saw elsewhere; they were kept for sale by old Waxler, the German basket maker, a most respected citizen, whose name I now learn was not Waxler but Weichsler. These dolls came in three sizes, the five-cent size was a midsummer favorite, because on its featureless head the blossoms of the Canterbury Bells fitted like a high azure cap. I can see rows of these wooden creatures sitting, thus crowned, stiffly around the trunk of the old Seckel Pear tree at a doll's tea-party.

By the constant trampling of our childish feet the earth at the end of the garden path was hard and smooth under the shadow of the Lilac trees near our garden fence; and this hard path, remote from wanderers in the garden, made a splendid plateau to use for flower balls. Once we fitted it up as a palace; circular walls of Balsam flowers set closely together shaped the ball-room. The dancers were blue and white Cantérbury Bells. Quadrilles were placed of little twigs, or strong flower stalks set firmly upright in the hard trodden earth, and on each of these a flower bell was hung so that the pretty reflexion of the scalloped edges of the corolla just touched the ground as the hooped petticoats swayed lightly in the wind.

We used to catch bumblebees in the Canterbury Bells, and hear them buzz and bump and tear their way out to liberty. We held the edges of the flower tightly pinched together, and were never stung. Besides its adaptability as a toy for children, the Canterbury Bell was beloved for its beauty in the garden. An appropriate folk name for it is Fair-in-sight. Healthy clumps grow tall and stately, towering up as high as childish heads; and the firm stalks are hung so closely in bloom. Nowadays

Hollyhocks in Garden of Kimball Homestead, Portsmouth, New Hampshire.

people plant expanses of Canterbury Bells; one at the beautiful garden at White Birches, Elmhurst, Illinois, is shown on page 111. I do not like this as well as the planting in our home garden when they are set in a mixed border, as shown opposite page 416. Our tastes in the flower world are largely influenced by what we were wonted to in childhood, not only in the selection of flowers, but in their placing in our gardens. The Canterbury Bell has historical interest through its being named for the bells borne by pilgrims to the shrine at Canterbury. I have been delighted to see plants of these sturdy garden favorites offered for sale of late years in New York streets in springtime, by street venders, who now show a tendency to throw aside Callas, Lilies, Tuberoses, and flowers of such ilk, and substitute shrubs and seedlings of hardy growth and satisfactory flowering. But it filled me with regret, to hear the pretty historic name — Canterbury Bells — changed in so short a residence in the city, by these Italian and German tongues to Gingerbread Bells — a sad debasement. Native New Englanders have seldom forgotten or altered an old flower name, and very rarely transferred it to another plant, even in two centuries of everyday usage. But I am glad to know that the flower will bloom in the flower pot or soap box in the dingy window of the city poor, or in the square foot of earth of the city squatter, even if it be called Gingerbread Bells.

I think we may safely affirm that the Hollyhock is the most popular, and most widely known, of all old-fashioned flowers. It is loved for its beauty, its associations, its adaptiveness. It is such a decorative flower, and looks of so much distinction in so many places. It is invaluable to the landscape gardener and to the architect; and might be named the wallflower, since it looks so well growing by every wall. I like it there, or by a fence-side, or in a corner, better than in the middle of flower beds. How many garden pictures have Hollyhocks? Sir Joshua Reynolds even used them as accessories of his portraits. They usually grow so well and bloom so freely. I have seen them in Connecticut growing wild — garden strays, standing up by ruined stone walls in a pasture with as much grace of grouping, as good form, as if they had been planted by our most skilful gardeners or architects. Many illustrations of them are given in this book; I need scarcely refer to them; opposite page 334 is shown a part of the four hundred stalks of rich bloom in a Portsmouth garden. There is a pretty semidouble Hollyhock with a single row of broad outer petals and a smaller double rosette for the centre; but the single flowers are far more effective. I like well the old single crimson flower, but the yellow ones are, I believe, the loveliest; a row of the yellow and white ones against an old brick wall is perfection. I can never repay to the Hollyhock the debt of gratitude I owe for the happy hours it furnished to me in my childhood. Its reflexed petals could be tied into such lovely silken-garbed dolls; its "cheeses" were one of the staple food supplies of our dolls' larder. I am sure in my childhood I would have warmly chosen the Hollyhock as my favorite flower.

The sixty-two folk names of the Foxglove give ample proof of its closeness to humanity; it is a familiar flower, a home flower. Of these many names I never heard but two in New England, and those but once; an old Irish gardener called the flowers Fairy Thimbles, and an English servant, Pops — this from the well-known habit of popping the petals on the palm of the hand. We used to build little columns of these Foxgloves by thrusting one within another, alternating purple and white; and we wore them for gloves, and placed them as foolscaps on the heads of tiny dolls. The beauty of the Foxglove in the garden is unquestioned; the spires of white bloom are, as Cotton Mather said of a pious and painful Puritan preacher, "a shining and white light in a golden candlestick improved for the sweet felicity of Mankind and to the honour of our Maker."

Opposite page 340 is a glimpse of a Box-edged garden in Worcester, whose blossoming has been a delight to me every summer of my entire life. In my childhood this home was that of flower-loving neighbors who had an established and constant system of exchange with my mother and other neighbors of flowers, plants, seeds, slips, and bulbs. The garden was serene with an atmosphere of worthy old age; you wondered how any man so old could so constantly plant, weed, prune, and hoe until you saw how he loved his flowers, and how his wife loved them. The Roses, Peonies, and Flower de Luce in this garden are sixty years old, and the Box also; the shrubs are almost trees. Nothing seems to be transplanted, yet all flourish; I suppose some plants must be pulled up, sometimes, else the garden would be a thicket. The varying grading of city streets has left this garden in a little valley sheltered from winds and open to the sun's rays. Here bloom Crocuses, Snowdrops, Grape Hyacinths, and sometimes Tulips, before any neighbor has a blossom and scarce a leaf. On a Sunday noon in April there are always flower lovers hanging over the low fences, and gazing at the welcome early blooms. Here if ever,

> "Winter, slumbering in the open air,
> Wears on his smiling face a dream of spring."

A close cloud of Box-scent hangs over this garden, even in midwinter; sometimes the Box edgings grow until no one can walk between; then drastic measures have to be taken, and the rows look ragged for a time.

I think much of my love of Box comes from happy associations with this garden. I used to like to go there with my mother when she went on what the Japanese would call "garden-viewing" visits, for at the lower end of the garden was a small orchard of the finest playhouse Apple trees I ever climbed (and I have had much experience), and some large trees bearing little globular early Pears; and there were rows of bushes of golden "Honey-blob" Gooseberries. The Apple trees are there still, but the Gooseberry bushes are gone. I looked for them this summer eagerly, but in vain; I presume the berries would have been sour had I found them.

An Autumn Path in a Worcester Garden.

In many old New England gardens the close juxtaposition and even intermingling of vegetables and fruits with the flowers gave a sense of homely simplicity and usefulness which did not detract

Hollyhocks at Tudor Place.

from the garden's interest, and added much to the child's pleasure. At the lower end of the long flower border in our garden, grew "Mourning Brides," white, pale lavender, and purple brown in tint. They opened under the shadow of a row of

Gooseberry bushes. I seldom see Gooseberry bushes nowadays in any gardens, whether on farms or in nurseries; they seem to be an antiquated fruit.

I have in my memory many other customs of childhood in the garden; some of them I have told in my book *Child Life in Colonial Days*, and there are scores more which I have not recounted, but most of them were peculiar to my own fanciful childhood, and I will not recount them here.

One of the most exquisite of Mrs. Browning's poems is *The Lost Bower;* it is endeared to me because it expresses so fully a childish bereavement of my own, for I have a lost garden. Somewhere, in my childhood, I saw this beautiful garden, filled with radiant blossoms, rich with fruit and berries, set with beehives, rabbit hutches, and a dove cote, and enclosed about with hedges; and through it ran a purling brook — a thing I ever longed for in my home garden. All one happy summer afternoon I played in it, and gathered from its beds and borders at will — and I have never seen it since. When I was still a child I used to ask to return to it, but no one seemed to understand; and when I was grown I asked where it was, describing it in every detail, and the only answer was that it was a dream, I had never seen and played in such a garden. This lost garden has become to me an emblem, as was the lost bower to Mrs. Browning, of the losses of life; but I did not lose all; while memory lasts I shall ever possess the happiness of my childhood passed in our home garden.

An Old Worcester Garden.

CHAPTER XVI

MEETIN' SEED AND SABBATH DAY POSIES

> "I touched a thought, I know
> Has tantalized me many times.
> Help me to hold it! First it left
> The yellowing Fennel run to seed."
>
> —ROBERT BROWNING.

Y "thought" is the association of certain flowers with Sunday; the fact that special flowers and leaves and seeds, Fennel, Dill, and Southernwood, were held to be fitting and meet to carry to the Sunday service. "Help me to hold it"—to record those simple customs of the country-side ere they are forgotten.

In the herb garden grew three free-growing plants, all three called indifferently in country tongue, "meetin' seed." They were Fennel, Dill, and Caraway, and similar in growth and seed. Caraway is shown on page 342. Their name was given because, in summer days of years gone by, nearly every woman and child carried to "meeting" on Sundays, bunches of the ripe seeds of one or all of these three plants, to nibble throughout the long prayers and sermon.

It is fancied that these herbs were anti-soporific, but I find no record of such power. On the contrary, Galen says Dill "procureth sleep, wherefore garlands of Dill are worn at feasts." A far more probable reason for its presence at church was the quality assigned to it by Pliny and other herbalists down to Gerarde, that of staying the "yeox or hicket or hicquet," otherwise the hiccough. If we can judge by the manifold remedies offered to allay this affliction, it was certainly very prevalent in ancient times. Cotton Mather wrote a bulky medical treatise entitled *The Angel of Bethesda.* It was never printed; the manuscript is owned by the American Antiquarian Society. The character of this medico-religious book may be judged by this opening sentence of his chapter on the hiccough: —

Caraway.

"The Hiccough or the Hicox rather, for it's a Teutonic word that signifies to sob, appears a Lively Emblem of the battle between the Flesh and the Spirit in the Life of Piety. The Conflict in the Pious Mind gives all the Trouble and same uneasiness as Hickox. Death puts an end to the Conflict."

Parson Mather gives Tansy and Caraway as remedies for the hiccough, but far better still — spiders, prepared in various odious ways; I prefer Dill.

Peter Parley said that "a sprig of Fennel was the theological smelling-bottle of the tender sex, and not unfrequently of the men, who from long sitting in the sanctuary, after a week of labor in the field, found themselves tempted to sleep, would sometimes borrow a sprig of Fennel, to exorcise the fiend that threatened their spiritual welfare."

Old-fashioned folk kept up a constant nibbling in church, not only of these three seeds, but of bits of Cinnamon or Lovage root, or, more commonly still, the roots of Sweet Flag. Many children went to brooksides and the banks of ponds to gather these roots. This pleasure was denied to us, but we had a Flag root purveyor, our milkman's daughter. This milkman, who lived on a lonely farm, used often to take with him on his daily rounds his little daughter. She sat with him on the front seat of his queer cart in summer and his queerer pung in winter, an odd little figure, with a face of gypsylike beauty which could scarcely be seen in the depths of the Shaker sunbonnet or pumpkin hood. If my mother chanced to see her, she gave the child an orange, or a few figs, or some little cakes, or almonds and raisins; in return the child would throw out to us violently roots of Sweet Flag, Wild Ginger, Snakeroot, Sassafras, and Apples or Pears, which she carried in a deep detached pocket at her side. She never spoke, and the milkman confided to my mother that he "took her around because she was so wild," by which he meant timid. We were firmly convinced that the child could not walk nor speak, and had no ears; and we were much surprised when she walked down the aisle of our church one Sunday as actively as any child could, displaying very natural ears. Her father had bought a home in the town that she might go to school. He was rewarded by her development into one of those scholars of phenomenal brilliancy, such as are occasionally produced from New England farmers' families. She also became a beauty of most unusual type. At her father's death she "went West." I have always expected to read of her as of marked life in some way, but I never have. Of course her family name may have been changed by marriage; but her Christian name, Appoline, was so unusual I could certainly trace her. If my wild and beautiful little milk girl reads these lines, I hope she will forgive me, for she certainly was queer.

Sun-dial of Jonathan Fairbanks.

When her residence was in town, Appoline did

not cease her gifts of country treasures. She brought on spring Sundays a very delightful addition to our Sabbath day nibblings and browsings, the most delicious mouthful of all the treasures of New England woods, what we called Pippins, the first tender leaves of the aromatic Checkerberry. In the autumn the spicy berries of the same plant filled many a paper cornucopia which was secretly conveyed to us.

It was also a universal custom among the elder folk to carry a Sunday posy; the stems were discreetly enwrapped with the folded handkerchief which also concealed the sprig of Fennel. Dean Hole tells us that a sprig of Southernwood was always seen in the Sunday smocks of English farm folk. Mary Howitt, in her poem, *The Poor Man's Garden*, has this verse: —

> "And here on Sabbath mornings
> The goodman comes to get
> His Sunday nosegay — Moss Rose bud,
> White Pink, and Mignonette."

This shows to me that the church posy was just as common in England as in America; in domestic and social customs we can never disassociate ourselves from England; our ways, our deeds, are all English.

Thoreau noted with pleasure when, at the last of June, the young men of Concord "walked slowly and soberly to church, in their best clothes, each with a Pond Lily in his hand or bosom, with as long a stem as he could get." And he adds thereto almost the only decorous and conventional picture he gives of himself, that he used in early life to go thus to church, smelling a Pond Lily, "its odor contrasting with and atoning for that of the sermon." He associated this universal bearing of the Lily with a very natural act, that of the first spring swim and bath, and pictured with delight the quiet Sabbath stillness and the pure opening flowers. He said the flower had become typical to him equally of a Sunday morning swim and of church-going. He adds that the young women carried on this floral Sunday, as a companion flower, their first Rose.

Bronze Sun-dial on Dutch Reformed Church, West End Avenue, New York.

This Sabbath bearing of the early Water Lilies may have been a local custom; a few miles from Walden Pond and Concord an old kinsman of mine throughout his long life (which closed twenty years ago) carried Water Lilies on summer Sundays to church; and starting with neighborly intent a short time before the usual hour of church service, he placed a single beautiful Lily in the pew of each of his old friends. All knew who was the flower bearer, and gentle smiles and nods of thanks would radiate across the old church to him. These lilies were gathered for him freshly each Sabbath morning by the young men of his family, who, as Thoreau tells, all took their morning bath in the pond throughout the summer.

Sun-dial on Boulder, Swiftwater, Pennsylvania.

There were conventions in these Sunday posies. I never heard of carrying sprays of Lemon Verbena or Rose Geranium, or any of the strong-scented herbs of the Mint family; but throughout eastern Massachusetts, especially in Concord and Wayland, a favorite posy was a spray of the refreshing, soft-textured leaves from what country folk called the Tongue plant—which was none other than Costmary, also called Beaver tongue, and Patagonian mint. As there has been recently much interest and discussion anent this

Tongue plant, I here give its botanical name *Chrysanthemum balsamita*, var. *tanacetoides*. A far more popular Sunday posy than any blossom was a sprig of Southernwood, known also everywhere as Lad's-love, and occasionally as Old Man and Kiss-me-quick-and-go. It was also termed Meeting plant from this universal Sunday use.

A restless little child was once handed during the church services in summer a bunch of Caraway seeds, and a goodly sprig of Southernwood. The little girl's mother listened earnestly to the long sermon, and was horrified at its close to find that her child had eaten the entire bunch of Caraway, stems and seeds, and all the bitter Southernwood. She was hurried out of church to the village doctor's, and spent a very unhappy hour or two as the result of her Nebuchadnezzar-like gorging.

Like many New Englanders, I dearly love the scent of Southernwood: —

> "I'll give to him
> Who gathers me, more sweetness than he knows
> Without me — more than any Lily could,
> I, that am flowerless, being Southernwood."

Southernwood bears a balmier breath than is ever borne by many blossoms, for it is sweet with the fragrance of memory. The scent that has been loved for centuries, the leaves that have been pressed to the hearts of fair maids, as they questioned of love, are indeed endeared.

Southernwood was a plant of vast powers. It was named in the fourteenth century as potent to

Buckthorn Arch in an Old Salem Garden.

cure talking in sleep, and other "vanityes of the heade." An old Salem sea captain had this recipe for baldness: "Take a quantitye of Suthernwoode and put it upon kindled coale to burn and being made into a powder mix it with oyl of radiches, and anoynt a bald head and you shall see great experiences." The lying old *Dispensatory* of Culpepper gave a rule to mix the ashes of Southernwood with "Old Sallet Oyl" which "helpeth those that are hair-fallen and bald."

Sun-dial at Emery Place, Brightwood, District of Columbia.

Far pleasanter were the uses of the plant as a love charm. Pliny did not disdain to counsel putting Southernwood under the pillow to make one dream of a lover. A sprig of Southernwood in an unmarried girl's shoe would bring to her the sight of her husband-to-be before night.

Sixty years ago two young country folk of New England were married. The twain built them a house and established their home. Since a sprig of Southernwood had played a romantic part in their courtship, each planted a bush at the side of the

Sun-dial at Traveller's Rest.

broad doorstone; and the husband, William, often thrust a bit of this Lad's-love from the flourishing bushes in the buttonhole of his woollen shirt, for he fancied the fresh scent of the leaves.

The twain had no children, and perhaps therefrom grew and increased in Hetty a fairly passionate love of exact order and neatness in her home — a trait which is not so common in New England housewives as many fancy, and which does not always find equal growth and encouragement in New England husbands. William chafed under the frequent and bitter reproofs for the muddy shoes, dusty garments, hanging straws and seeds which he brought into his wife's orderly paradise, and the jarring culminated one night over such a trifle, a green sprig of Lad's-love which he had dropped and trodden into the freshly washed floor of the kitchen, where it left a green stain on the spotless boards.

The quarrel flamed high, and was followed by an ominous calm which was not broken at breakfast. It would be impossible to express in words Hetty's emotions when she crossed her threshold to set her shining milk tins in the morning sunlight, and saw on one side of the doorstone a yawning hole where had grown for ten years William's bunch of Lad's-love. He had driven to the next village to sell some grain, so she could search unseen for the vanished emblem of domestic felicity, and soon she found it, in the ditch by the public road, already withered in the hot sun.

When her husband went at nightfall to feed and water his cattle, he found the other bush of Lad's-love, which had been planted with such affectionate sentiment, trodden in the mire of the pigpen, under the feet of the swine.

They lived together for thirty years after this

crowning indignity. The grass grew green over the empty holes by the doorside, but he never forgave her, and they never spoke to each other save in direst necessity, and then in fewest words. Yet they were not wicked folk. She cared for his father and mother in the last years of their life with a devotion that was fairly pathetic when it was seen that the old man was untidy to a degree, and absolutely oblivious of all her orderly ways and wishes. At their death he sent for and "homed," as the expression ran, a brother of hers who was almost blind, and paid the expenses of her nephew through college — but he died unforgiving; the sight of that beloved Southernwood — in the pigpen — forever killed his affection.

CHAPTER XVII

SUN-DIALS

"'Tis an old dial, dark with many a stain,
In summer crowned with drifting orchard bloom,
Tricked in the autumn with the yellow rain,
And white in winter like a marble tomb.

"And round about its gray, time-eaten brow
Lean letters speak — a worn and shattered row: —
'I am a Shade; A Shadowe too arte thou;
I mark the Time; saye, Gossip, dost thou soe?'"

— AUSTIN DOBSON.

CENTURY or more ago, in the heart of nearly all English gardens, and in the gardens of our American colonies as well, there might be seen a pedestal of varying material, shape, and pretension, surmounted by the most interesting furnishing in "dead-works" of the garden, a sun-dial. In public squares, on the walls of public buildings, on bridges, and by the side of the way, other and simpler dials were found. On the walls of country houses and churches vertical sun-dials were displayed; every English town held them by scores. In Scotland, and to some extent in England, these sun-dials still are found; in fine old gardens the most richly carved dials are standing; but in America they have become so rare that many people have never seen one. In many of the formal gardens planned by our skilled architects, sun-dials

Two Old Cronies, the Sun-dial and Bee Skepe.

are now springing afresh like mushroom growth of a single night, and some are objects of the greatest beauty and interest.

If the claims of antiquity and historical association have aught to charm us, every sun-dial must be assured of our interest. The most primitive mode of knowing of the midday hour was by a "noon mark," a groove cut or line drawn on door or window sill which indicated the meridian hour through a shadow thrown on this noon mark. A good guess as to the hours near noon could be made by noting the distance of the shadow from the noon mark. I chanced to be near an old noon mark this summer as the sun warned that noon approached; I noted that the marking shadow crossed the line at twenty minutes before noon by our watches—which, I suppose, was near enough to satisfy our "early to rise" ancestors. Meridian lines were often traced with exactness on the floors of churches in Continental Europe.

An advance step in accuracy and elegance was made when a simple metal sun-dial was affixed to the window sill instead of cutting the rude noon mark. Soon the sun-dial was set on a simple pedestal near the kitchen window, so that the active worker within might glance at the dial face without ceasing in her task. Such a sun-dial is shown on page 354, as it stands under the "buttery" window cosily hobnobbing with its old crony of many years, the bee skepe. One could wish to be a bee, and live in that snug home under the Syringa bush.

Portable sun-dials succeeded fixed dials; they have been known as long as the Christian era; shepherds' dials were the "Kalendars" or "Cylindres" about which treatises were written as early as the thirteenth century. They were small cylinders of wood or ivory, having at the top a kind of stopper with a hinged gnomon; they are still used in the

Pyrenees. Pretty little "ring-dials" of brass, gold, or silver, are constructed on the same principle. The exquisitely wrought portable dial shown on this page is a very fine piece of workmanship, and must have been costly. It is dated 1764, and is eleven inches in diameter. It is a perfect example

Portable Sun-dial.

of the advanced type of dial made in Italy, which had a simpler form as early certainly as A.D. 300. The compass was added in the thirteenth century. The compass-needle is missing on this dial, its only blemish. The Italians excelled in dial-making; among their interesting forms were the cross-shaped dials evidently a reliquary.

Portable dials were used instead of watches. There is at the Washington headquarters at Morristown a delicately wrought oval silver case, with compass and sun-dial, which was carried by one of the French officers who came here with Lafayette; George Washington owned and carried one.

The colonists came here from a land set with dials, whether they sailed from Holland or England. Charles I had a vast fancy for dials, and had them placed everywhere; the finest and most curious was the splendid master dial placed in his private gardens at Whitehall; this had five dials set in the upper part, four in the four corners, and a great horizontal concave dial; among these were scattered equinoctial dials, vertical dials, declining dials, polar dials, plane dials, cylindrical dials, triangular dials; each was inscribed with explanatory verses in Latin. Equally beautiful and intricate were the dials of Charles II, the most marvellous being the vast pyramid dial bearing 271 different dial faces.

Those who wish to learn of English sun-dials should read Mrs. Gatty's *Book of Sun-dials*, a massive and fascinating volume. No such extended record could be made of American sun-dials; but it pleases me that I know of over two hundred sun-dials in America, chiefly old ones; that I have photographs of many of them; that I have copies of many hundred dial mottoes, and also a very fair collection of the old dial faces, of various metals and sizes.

I know of no public collection of sun-dials in America save that in the Smithsonian Institution, and that is not a large one. Several of our Historical Societies own single sun-dials. In the Essex Institute is the sun-dial of Governor Endicott; another, shown on page 344, was once the property of my far-away grandfather, Jonathan Fairbanks; it is in the Dedham Historical Society.

All forms of sun-dials are interesting. A simple but accurate one was set on Robins Island by the

Sun-dial in Garden of Frederick J. Kingsbury, Esq.

late Samuel Bowne Duryea, Esq., of Brooklyn. Taking the flagpole of the club house as a stylus, he laid the lines and figures of the dial-face with small dark stones on a ground of light-hued stones, all set firmly in the earth at the base of the pole. Thus was formed, with the simplest materials, by one who ever strove to give pleasure and stimulate knowledge in all around him, an object which not only told the time o' the day, but afforded gratification, elicited investigation, and awakened sentiment in all who beheld it.

A similar use of a vertical pole as a primitive gnomon for a sun-dial seems to have been common to many uncivilized peoples. In upper Egypt the natives set up a palm rod in open ground, and arrange a circle of stones or pegs around it, calling it an *alka*, and thus mark the hours. The ploughman leaves his buffalo standing in the furrow while he learns the progress of time from this simple dial — and we recall the words of Job, "As a servant earnestly desireth a shadow."

Sun-dial at Morristown, New Jersey.

The Labrador Indians, when on the hunt or the march, set an upright stick or spear in the snow, and draw the line of the shadow thus cast. They then stalk on their way; and the women, heavily laden with provisions, shelter, and fuel, come slowly along two or three hours later, note the distance between the present shadow and the line drawn by their lords, and know at once whether they must gather up the stick or spear and hurry along, or can rest for a short time on their weary march. This is a primitive but exact chronometer.

There are serious objections to quoting from Charles Lamb: you are never willing to end the transcription—you long to add just one phrase, one clause more. Then, too, the purity of the pearl which you choose seems to render duller than their wont the leaden sentences with which you enclose it as a setting. Still, who could write of sun-dials without choosing to transcribe these words of Lamb's?

"What a dead thing is a clock, with its ponderous embowelments of lead or brass, its pert or solemn dulness of communication, compared with the simple altar-like structure and silent heart-language of the old dial! It stood as the garden god of Christian gardens. Why is it almost everywhere banished? If its business use be suspended by more elaborate inventions, its moral uses, its beauty, might have pleaded for its continuance. It spoke of moderate labors, of pleasures not protracted after sunset, of temperance and good hours. It was the primitive clock, the horologe of the first world. Adam could scarce have missed it in Paradise. The 'shepherd carved it out quaintly in the sun,' and turning philosopher by the very occupation, provided it with mottoes more touching than tombstones."

Sun-dial mottoes still can be gathered by hundreds; and they are one record of a force in the development of our literate people. For it was long after we had printing ere we had any general class of folk, who, if they could read, read anything save the Bible. To many the knowledge of reading came from the deciphering of what has been happily termed the Literature of the Bookless. This literature was placed that he who ran might read; and its opening chapters were in the form of inscriptions and legends and mottoes which were placed, not only on buildings and walls, and pillars and bridges, but on household furniture and table utensils.

Yes, Toby! It's Three O'clock.

The inscribing of mottoes on sun-dials appears to have sprung up with dial-making; and where could a strict moral lesson, a suggestive or inspiring thought, be better placed? Even the most heedless or indifferent passer-by, or the unwilling reader could not fail to see the instructive words when he cast his glance to learn the time.

The mottoes were frequently in Latin, a few in Greek or Hebrew; but the old English mottoes seem the most appealing.

ABUSE ME NOT I DO NO ILL
I STAND TO SERVE THEE WITH GOOD WILL
AS CAREFUL THEN BE SURE THOU BE
TO SERVE THY GOD AS I SERVE THEE.

A CLOCK THE TIME MAY WRONGLY TELL
I NEVER IF THE SUN SHINE WELL.

AS A SHADOW SUCH IS LIFE.

I COUNT NONE BUT SUNNY HOURS.

BE THE DAY WEARY, BE THE DAY LONG
SOON IT SHALL RING TO EVEN SONG.

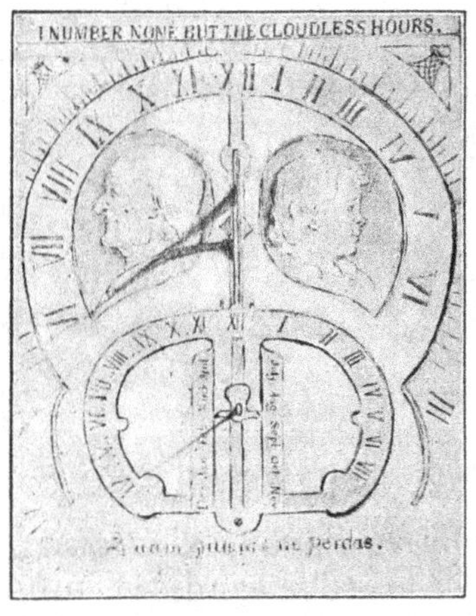

Face of Dial at Sag Harbor, Long Island.

Scriptural verses have ever been favorites, especially passages from the Psalms: "Man is like a thing of nought, his time passeth away like a shadow." "My time is in Thy hand." "Put not off from day to day." "Oh, remember how short my time is." Some of the Latin mottoes are very beautiful.

Poets have written special verses for sundials. These noble lines are by Walter Savage Landor: —

IN HIS OWN IMAGE THE CREATOR MADE,
HIS OWN PURE SUNBEAM QUICKENED THEE, O MAN!
THOU BREATHING DIAL! SINCE THE DAY BEGAN
THE PRESENT HOUR WAS EVER MARKED WITH SHADE.

The motto, *Horas non numero nisi serenas*, in various forms and languages, has ever been a favorite. From an old album I have received this poem written by Professor S. F. B. Morse; there is a note with it in Professor Morse's handwriting, saying he saw the motto on a sun-dial at Worms:—

TO A. G. E.

Horas non numero nisi serenas.

The sun when it shines in a clear cloudless sky
  Marks the time on my disk in figures of light;
If clouds gather o'er me, unheeded they fly,
  I note not the hours except they be bright.

So when I review all the scenes that have past
  Between me and thee, be they dark, be they light,
I forget what was dark, the light I hold fast;
  I note not the hours except they be bright.

SAMUEL F. B. MORSE,
Washington, March, 1845.

The sun-dial seems too classic an object, and too serious a teacher, to bear a jesting motto. This sober pun was often seen:—

LIFE'S BUT A SHADOWE
MAN'S BUT DUST
THIS DYALL SAYES
DY ALL WE MUST.

The sun-dial does not lure to "idle dalliance." Nine-tenths of the sun-dial mottoes tersely warn you not to linger, to haste away, that time is fleeting, and your hours are numbered, and therefore to "be about your business." In a single moment and at a single glance the sun-dial has said its lesson, has told its absolute message, and there is no reason for you to gaze at it longer. Its very position, too, in the unshaded rays of the sun, does not invite you to long companionship, as do the shady lengths of a pergola, or a green orchard seat. Still, I would ever have a gar-

Sun-dial in Garden of Grace Church Rectory, New York.

den seat near a sun-dial, especially when it is a work of art to be studied, and with mottoes to be remembered. For even in hurrying America the sun-dial seems — like a guide-post — a half-human thing, for which we can feel an almost personal interest.

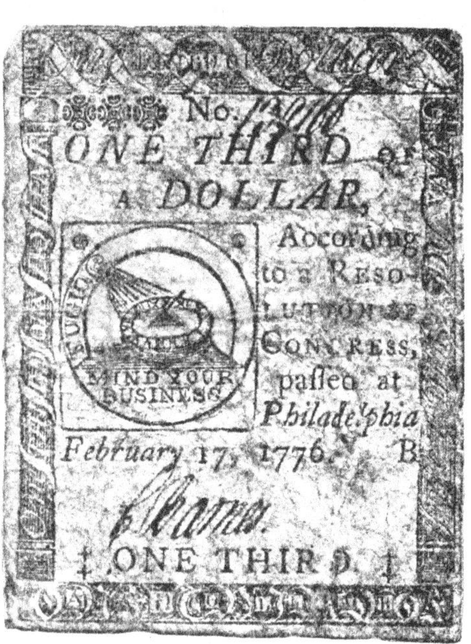

Fugio Bank-note.

The figure of a sun-dial played an interesting part in the early history of the United States. In the first set of notes issued for currency by the American Congress was one for the value of one third of a dollar. One side has the chain of links bearing the names of the thirteen states, enclosing a sunburst bearing the words, *American Congress, We are One.* The reverse side is shown on this page. It bears a print of a sun-dial, with the motto, *Fugio, Mind Your Business.* The so-called "Franklin cent" has a similar design of a sun-dial with the same motto, and there was a beautiful "Fugio dollar" cast in silver, bronze, and pewter. Though this design and motto were evidently Franklin's taste, the motto in its use on a sun-dial was not original with Franklin, nor with any one else in the Congress, for it had been seen on dials on many English churches and houses. In the form, "Begone about Your Business," it was on a house in the Inner Temple; this is the tradition of the origin of this motto. The dialler sent for a motto to place under the dial, as he had been instructed by the Benchers; when the man arrived at the Library, he found but one surly old gentleman poring over a musty book. To him he said, "Please, sir, the gentlemen told me to call this hour for a motto for the sun-dial." "Begone about your business," was the testy answer. So the man painted the words under the dial; and the chance words seemed so appropriate to the Benchers that they were never removed. It is told of Dean Cotton of Bangor that he had a cross old gardener who always warded off unwelcome visitors to the deanery by saying to every one who approached, "Go about your business!" After the gardener's death the dean had this motto engraved around the sun-dial in the garden, "Goa bou tyo urb us in ess, 1838." Thus the gardener's growl became his epitaph. Another form was, "Be about Your Business," and it is a suggestive fact that it was on a dial on the General Post-office in London in 1756. Franklin's interest in and knowledge of postal matters, his long residence in London, and service under the crown as American postmaster general, must have familiarized him with this dial, and I am convinced it furnished to him the notion for the design on the first bank-note and coins of the new nation.

Sun-dial at "Washington House," Little Brington, England.

An interesting bit of history allied to America is given to us in the finding of a sun-dial which gives to American students of heraldic antiquities another dated shield of the Washington "stars and stripes."

In Little Brington, Northamptonshire, stands a house known as "The Washington House," which gave shelter to the Washingtons of Sulgrave after the fall of their fortunes. Within a stone's throw of the house has recently been found a sun-dial having the Washington arms (argent) two bars, and in chief three mullets (gules) carved upon it, with the date 1617. The existence of this stone has been known for forty years; but it has never been closely examined and noted till recently. It is a circular slab of sandstone three inches thick and sixteen inches in diameter. The gnomon is lacking. The lines, figures, and shield are incised, and the letters R. W. can be dimly seen. These were probably

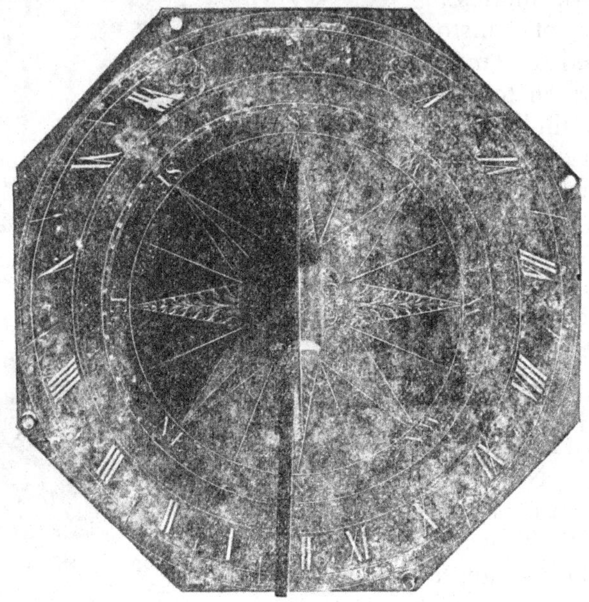

Dial-face from Mount Vernon.

the initials of Robert Washington, great uncle of the two emigrants to Virginia.

Through the kindness of Mr. A. L. Y. Morley, a faithful antiquary of Great Barrington, I have the pleasure of giving, on page 367, a representation of this interesting dial. It is shown leaning against the "pump-stand" in the yard of the "Washington House"; and the pump seems as ancient as the dial.

In this book are three other sun-dials associated with George Washington. At Mount Vernon there stands at the front of the entrance door a modern sun-dial. The fine old metal dial-face, about ten inches in diameter, which in Washington's day was placed on the same site, is now the property of Mr. William F. Havemeyer, Jr., of New York. It was given to him by Mr. Custis; a picture of it is shown on page 368. This dial-face is a splendid relic; one closely associated with Washington's everyday life, and full of suggestion and sentiment to every thoughtful beholder. The sun-dial which stood in the old Fredericksburg garden of Mary Washington,

Sun-dial of Mary Washington, Fredericksburg, Virginia.

the mother of George Washington, still stands in Fredericksburg, in the grounds of Mr. Doswell. A photograph of it is reproduced on page 369. The fourth historic dial is on page 371. It is the one at Kenmore, the home built by Fielding Lewis for his bride, Betty Washington, the sister of George Washington, on ground adjoining her mother's home. A part of the garden which connected these two Washington homes is shown on page 228. These three American sun-dials afford an interesting proof of the universal presence of sun-dials in Virginian homes of wealth, and they also show the kind of dial-face which was generally used. Another ancient dial (page 350) at Travellers' Rest, a near-by Virginian country seat, is similar in shape to these three, and differs but little in mounting.

In Pennsylvania and Virginia sun-dials have lingered in use in front of court-houses, on churches, and in a few old garden dials. In New England I scarcely know an old garden dial still standing in its original place on its original pedestal. Four old ones of brass or pewter are shown in the illustration on page 379. These once stood in New England gardens or on the window sills of old houses; one was taken from a sunny window ledge to give to me.

Perhaps the attention paid the doings of the American Philosophical Society, and the number of scientists living near Philadelphia, may account for the many sun-dials set up in the vicinity of the town. Godfrey, the maker of Godfrey's Quadrant, was one of those scientific investigators, and must have been a famous "dialler."

On page 373 is shown an ancient sun-dial in the garden of Charles F. Jenkins, Esq., in Germantown, Pennsylvania. This sun-dial originally belonged to Nathan Spencer, who lived in Germantown

Kenmore, the Home of Betty Washington Lewis.

prior to and during the Revolutionary War. Hepzibah Spencer, his daughter, married, and took the sun-dial to Byberry. Her daughter carried the sun-dial to Gwynedd when her name was changed to Jenkins; and their grandson, the present owner, rescued it from the chicken house with the gnomon missing, which was afterward found. Its inscription, "Time waits for No Man," is an old punning device on the word gnomon.

At one time dialling was taught by many a country schoolmaster, and excellent and accurate sun-dials were made and set up by country workmen, usually masons of slight education. In Scotland the making of sun-dials has never died out. In America many pewter sun-dials were cast in moulds of steatite or other material. A few dial-makers still remain; one in lower New York makes very interesting-looking sun-dials of brass, which, properly discolored and stained, find a ready sale in uptown shops. I doubt if these are ever made for any special geographical point, but there is in a small Pennsylvania town an old Quaker who makes carefully calculated and accurate sun-dials, computed by logarithms for special places. I should like to see him "sit like a shepherd carving out dials, quaintly point by point." I have a very pretty circular brass dial of his making, about eight inches in diameter. He writes me that "the dial sent thee is a good students' dial, fit to set outside the window for a young man to use and study by in college," which would indicate to me that my Quaker dialler knows another type of collegian from those of my acquaintance, who would find the time set by a sun-dial rather slow.

There have been those who truly loved sun-dials. Sir William Temple ordered that after his death his heart should be buried under the sun-dial in his garden — where his heart had been in life. 'Tis not unusual to see a sun-dial over the gate to a burial ground, and a noble emblem it is in that place; one at Mount Auburn Cemetery,

Sun-dial in Garden of Charles F. Jenkins, Esq., Germantown, Pennsylvania.

near Boston, bears a pleasing motto written originally by John G. Whittier for his friend, Dr. Henry Ingersoll Bowditch, and inscribed on a beautiful silver sun-dial now owned by Dr. Vincent Y. Bowditch of Boston, Massachusetts. A facsimile of this dial was also placed before the Manor House on the island of Naushon by Mr. John M. Forbes in memory of Dr. Bowditch. The lines run thus: —

WITH WARNING HAND I MARK TIME'S RAPID FLIGHT
FROM LIFE'S GLAD MORNING TO ITS SOLEMN NIGHT.
YET, THROUGH THE DEAR GOD'S LOVE I ALSO SHOW
THERE'S LIGHT ABOVE ME, BY THE SHADE BELOW.

A sun-dial is to me, in many places, a far more inspiring memorial than a monument or tablet. Let me give as an example the fine sun-dial, designed by W. Gedney Beatty, Esq., and shown on page 359, which was erected on the grounds of the Memorial Hospital at Morristown, New Jersey, by the Society of the Daughters of the American Revolution, to mark the spot where Washington partook of the Communion.

What dignified and appropriate church appointments sun-dials are. A simple and impressive bronze vertical dial on the wall of the Dutch Reformed Church on West End Avenue, New York, is shown on page 346. The sun-dial standing before the rectory of Grace Church on Broadway, New York, is on page 364.

There is ever much question as to a suitable pedestal for garden sun-dials: it must not stand so high that the dial-face cannot be looked down upon by grown persons; it must not be so light as to seem rickety, nor so heavy as to be clumsy. A very good rule is to err on the side of simplicity in sun-dials for ordinary gardens. What I regard

Sun-dial at Ophir Farm, White Plains, New York, Country-seat of Hon. Whitelaw Reid.

as a very satisfactory pedestal and mounting in every particular may be seen in the illustration facing page 80, showing the sun-dial in the garden of Charles E. Mather, Esq., at Avonwood

Court, Haverford, Pennsylvania. Sometimes the pillars of old balustrades, old fence posts, and even parts of old tombs and monuments, have been used as pedestals for sun-dials. How pleasantly Sylvana in her *Letters to an Unknown Friend*, tells us and shows to us her cheerful sun-dial mounted on the four corners of an old tombstone with this fine motto cut into the upper step, *Lux et umbra vicissim sed semper amor*. I mean to search the stone-cutters' waste heap this summer and see whether I cannot rob the grave to mark the hours of my life. Charles Dickens had at Gadshill a sun-dial set on one of the pillars of the balustrade of Old Rochester Bridge. From Italy and Greece marble pillars have been sent from ancient ruins to be set up as dial pedestals.

If possible, the pedestal as well as the dial-face of a handsome sun-dial should have some significance through association, suggestion, or history. At Ophir Farm, White Plains, New York, the country-seat of Hon. Whitelaw Reid, may be seen a sun-dial full of exquisite significance. It is shown on page 375. The signs of the Zodiac in finely designed bronze are set on the symmetrical marble pedestal, and seem wonderfully harmonious and appropriate. This sun-dial is a literal exemplification of the words of Emerson: —

"A calendar
Exact to days, exact to hours,
Counted on the spacious dial
Yon broidered Zodiac girds."

The dial-face is upheld by a carefully modelled tortoise in bronze, which is an equally suggestive emblem, connected with the tradition, folk-lore, and religious beliefs of both primitive and cultured peoples; it is specially full of meaning in this place. The whole sun-dial shows much thought and æsthetic perception in the designer and owner, and cannot fail to prove gratifying to all observers having either sensibility or judgment.

Occasionally a very unusual and beautiful sun-dial standard may be seen, like the one in the Rose garden at Yaddo, Saratoga, New York, a copy of rarely beautiful Pompeian carvings. A representation of this is shown on page 86. Copies of simpler antique carvings make excellent sun-dial pedestals; a safe rule to follow is to have a reproduction made of some well-proportioned English or Scotch pedestal. The latter are well suited to small gardens. I have drawings of several Scotch sun-dials and pedestals which would be charming in American gardens. In the gardens at Hillside, by the side of the Shakespeare Border is a sun-dial (page 378) which is an exact reproduction of the one in the garden at Abbotsford, the home of Sir Walter Scott. This pedestal is suited to its surroundings, is well proportioned; and has historic interest. It forms an excellent example of Charles Lamb's "garden-altar."

On a lawn or in any suitable spot the dial-face can be mounted on a boulder; one is here shown. I prefer a pedestal. For gardens of limited size, much simplicity of design is more pleasing and more fitting than any elaborate carving. In an Italian garden, or in any formal garden whose work in stone or marble is costly and artistic, the sun-dial pedestal should be

Sun-dial at Hillside, Menand's, near Albany, New York.

the climax in richness of carving of all the garden furnishing. I like the pedestal set on a little platform, so two or three steps may be taken up to it from the garden level; but after all, no rules can be given for the dial's setting. It may be planted with vines, or stand unornamented; it may be set low, and be looked down upon, or it may be raised high up on a side wall; but wherever it is, it must not be for a single minute in shadow; no trees or overhanging shrubs should be near it; it is a child of the sun, and lives only in the sun's full rays.

In the lovely old garden at the home of Frederick J. Kingsbury, Esq., at Waterbury, Conn., is a

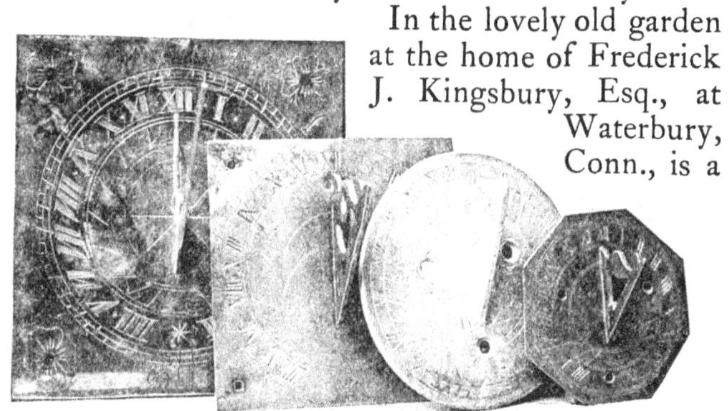

Old Brass and Pewter Dial-faces.

sun-dial bearing the motto, "*Horas non numero nisi serenas*," and the dates 1739–1751,—the dates of the building of the old and new houses on land that has been in the immediate family since 1739. Around this dial is a crescent-shaped bed of Zinnias, and very satisfactory do they prove. This garden has fine Box edgings; one is shown on page 173, a Box walk, set in 1851 with ancient Box brought from the garden of Mr. Kingsbury's great-great-grandfather.

The gnomon of a sun-dial is usually a simple

plate of metal in the general shape of a right-angled triangle, cut often in some pierced design, and occasionally inscribed with a motto, name, or date. Sometimes the dial-maker placed on the gnomon various Masonic symbols — the compass, square, and triangle, or the coat of arms of the dial owner.

One old English dial fitting we have never copied in America. It was the taste of the days of the Stuart kings, days of constant jesting and amusement and practical jokes. Concealed water jets were placed which wet the clothing of the unwary one who lingered to consult the dial-face.

The significance of the sun-dial, as well as its classicism, was sure to be felt by artists. In the paintings of Holbein, of Albert Dürer, dials may be seen, not idly painted, but with symbolic meaning. The mystic import of a sun-dial is shown in full effect in that perfect picture, *Beata Beatrix*, by Dante Gabriel Rossetti. I have chosen to show here (facing page 380) the *Beata Beatrix* owned by Charles L. Hutchinson, Esq., of Chicago, as being less photographed and known than the one of the British Gallery, from which it varies slightly and also because it has the beautiful predella. In this picture, in the words of its poet-painter: —

> "Love's Hour stands.
> Its eyes invisible
> Watch till the dial's thin brown shade
> Be born — yea, till the journeying line be laid
> Upon the point."

Beata Beatrix.

Andrew Marvell wrote two centuries ago of the floral sun-dials which were the height of the gardening mode of his day: —

"How well the skilful gardener drew
Of flowers and herbs this dial new.
When from above the milder sun
Does through a fragrant zodiac run;
And as it works the industrious bee
Computes its time as well as we!
How could such sweet and wholesome hours
Be reckoned but with herbs and flowers!"

These were sometimes set of diverse flowers, sometimes of Mallows. Two of growing Box are

The Faithful Gardener.

described and displayed in the chapter on Box edgings.

Linnæus made a list of forty-six flowers which constituted what he termed the Horologe or Watch of Flora, and he gave what he called their exact hours of rising and setting. He divided them into three classes: Meteoric, Tropical, and Equinoctial flowers. Among those which he named are: —

	Opening Hour.	Closing Hour.
Dandelion . . .	5–6 A.M.	8–9 P.M.
Mouse-ear Hawkweed .	8 A.M.	2 P.M.
Sow Thistle . .	5 A.M.	11–12 P.M.
Yellow Goat-beard .	3–5 A.M.	9–10 (?)
White Water Lily .	7 A.M.	7 P.M.
Day Lily . . .	5 A.M.	7–8 P.M.
Convolvulus . .	5–6 A.M.	
Mallow . . .	9–10 A.M.	
Pimpernel . . .	7–8 A.M.	
Portulaca . . .	9–10 A.M.	
Pink (*Dianthus prolifer*)	8 A.M.	1 P.M.
Succory . . .	4–5 A.M.	
Calendula . . .	7 A.M.	3–4 P.M.

Of course these hours would vary in this country. And I must say very frankly that I think we should always be behind time if we trusted to Flora's Horologe. This floral clock of Linnæus was calculated for Upsala, Sweden; De Candolle gave another for Paris, and one has been arranged for our Eastern states.

CHAPTER XVIII

GARDEN FURNISHINGS

"Furnished with whatever may make the place agreeable, melancholy, and country-like."

— *Forest Trees*, JOHN EVELYN, 1670.

UAINT old books of garden designers show us that much more was contained in a garden two centuries ago, than now; it had many more adjuncts, more furnishings; a very full list of them has been given by Batty Langley in his *New Principles of Gardening*, etc., 1728. Some seem amusing—as haystacks and woodpiles, which he terms "rural enrichments." Of water adornments there were to be purling streams, basins, canals, fountains, cascades, cold baths. There were to be aviaries, hare warrens, pheasant grounds, partridge grounds, dove-cotes, beehives, deer paddocks, sheep walks, cow pastures, and "manazeries" (menageries?); physic gardens, orchards, bowling-greens, hop gardens, orangeries, melon grounds, vineyards, parterres, fruit yards, nurseries, sun-dials, obelisks, statues, cabinets, etc., decorated the garden walks. There were to be land gradings of mounts, winding valleys, dales, terraces, slopes, borders, open plains, labyrinths, wildernesses, "serpentine meanders," "rude-coppices," precipices, amphitheatres. His "serpentine meanders" had large opening spaces at proper distances, in one of which might be placed a small fruit garden, a "cone of evergreens," or a "Paradice-Stocks,"—about which latter mysterious garden adornment I think we must be content to remain in ignorance, since he certainly has given us ample variety to choose from without it.

Other "landscapists" placed in their gardens old ruins, misshapen rocks, and even dead trees, in order to look "natural."

In 1608 Henry Ballard brought out *The Gardener's Labyrinth*—a pretty good book, shut away from the most of us by being printed in black letter. He says:—

> "The framing of sundry herbs delectable, with waies and allies artfully devised is an upright herbar."

Herbars, or arbors, were of two kinds: an upright arbor, which was merely a covered lean-to attached to a fence or wall; and a winding or "arch-arbor" standing alone. He names "archherbs," which are simply climbing vines to set "winding in arch-manner on withie poles." "Walker and sitters thereunder" are thereby comfortably protected from the heat of the sun. These upright arbors were in high favor; Ballard says they offered "fragrant savours, delectable sights, and sharpening of the memory."

Tree arbors were in use in Elizabethan times,

A Garden Lyre at Waterford, Virginia.

platforms built in the branches of large trees. Parkinson called one that would hold fifty men, "the goodliest spectacle that ever his eyes beheld." A distinction was made between arbors and bowers. The arbor might be round or square, and was domed over the top; while the long arched way was a bower. In our Southern states that special use of the word bower is still universal, especially in the term Rose bowers. A quaint and universal furnishing of old Southern gardens were the trellises known as garden lyres. Two are shown in this chapter, from Waterford, Virginia; one bearing little foliage and another embowered in vines, in order to show what a really good vine support they were. Garden lyres and Rose bowers are rotting on the ground in old Virginia gardens, and I fear they will never be replaced.

The word pergola was seldom heard here a century ago, save as used by the few who had travelled in Italy; but pergolas were to be found in many an old American garden. An ancient oval pergola still stands at Arlington, that beautiful spot which was once the home of the Virginia Lees, and is now the home of the honored dead of our Civil War. This old pergola has remained unharmed through fierce conflict, and is wreathed each spring with the verdure of vines of many kinds. It is twenty feet wide between the pillars, and forms an oval one hundred feet long and seventy wide, and when in full greenery is a lovely thing. It was called — indeed it is still termed in the South — a "green gallery," a word and thing of mediæval days.

There are many pretty trellises and vine supports and arbors which can be made of light poles and

A Virginia Lyre with Vines.

rails, but I do not like to hear the pretentious name, pergola, applied to them. A pergola must not be a mean, light-built affair. It should be of good proportions and substantial materials. It need not be made with brick or marble pillars; natural tree trunks of good size serve as well. It should look as if it had been built with care and stability, and that the vines had been planted and trained by skilled gardeners. A pergola may have a dilapidated Present and be endurable; but it should show evidences of a substantial Past.

Little sisters of the pergola are the *charmilles*, or bosquets, arches of growing trees, whose interlaced boughs have no supports of wood as have the pergolas. When these arches are carefully trained and pruned, and the ground underneath is laid with turf or gravel, they form a delightful shady walk.

Charming covered ways can be easily made by polling and training Plum or Willow trees. Arches are far too rare in American gardens. The few we have are generally old ones. In Mrs. Pierson's garden in Salem the splendid arch of Buckthorn is a hundred and twenty five years old. Similar ones are at Indian Hill. Cedar was an old choice for hedges and arches. It easily winter-kills at the base, and that is ample reason for its rejection and disuse.

The many garden seats of the old English garden were perhaps its chief feature in distinction from American garden furnishings to-day. In a letter written from Kenilworth in 1575 the writer told of garden seats where he sat in the heat of summer, "feeling the pleasant whisking wynde." I have walked through many a large modern garden in the summer heat, and longed in vain for a shaded seat from which to regard for a few moments the garden treasures and feel the whisking wind, and would gladly have made use of the temporary presence of a wheelbarrow.

Old Iron Gate at Westover-on-James.

Seats of marble and stone are in many of our modern formal gardens; a pretty one is in the garden at Avonwood Court.

Grottoes, arbors, and summer-houses were all of importance in those days, when in our latitude and climate men had not thought to build piazzas surrounding the house and shadowing all the ground floor rooms. We are beginning to think anew of the value of sunlight in the parlors and dining rooms of our summer homes, which for the past thirty or forty years have been so darkened by our wide piazzas. Now we have fewer piazzas and more peristyles, and soon we shall have summer-houses and garden houses also.

There are preserved in the South, in spite of war and earthquake, a number of fine examples of old wrought-iron garden gates. King William of England introduced these artistic gates into England, and they were the height of garden fashion. Among them were the beautiful gates still at Hampton Court, and those of Bulwich, Northamptonshire. They were called *clair-voyees* on account of the uninterrupted view they permitted to those without and within the walls. These were often painted blue; but in America they were more sober of tint, though portions were gilded. One of the old gates at Westover-on-James is here shown, and on page 390 the rich wrought-iron work in the courtyard at the home of Colonel Colt in Bristol, Rhode Island. This is as fine as the house, and that is a splendid example of the best work of the first years of the nineteenth century.

Fountains were seen usually in handsome gardens in the South; simple water jets falling in a handsome basin of marble or stone. Statuary of marble or lead was never common in old American gardens, though pretentious gardens had examples. To-day, in our carefully thought-out gardens, the garden statuary is a thing of beauty and often of meaning, as the figure shown on page 84. Usually our statues are of marble, sometimes a Japanese bronze is seen.

Iron-work in Court of Colt Mansion, Bristol, Rhode Island.

In the old black letter *Gardener's Labyrinth*, a very full description is given of old modes of watering a garden. There was a primitive and very limited system of irrigation, the water being raised by "well-swipes"; there were very handy puncheons, or tubs on wheels, which could be trundled down the garden walk. There was also a formidable "Great Squirt of Tin," which was said to take "mighty strength" to handle, and which looked like a small cannon; with it was an ingenious bent tube of tin by which the water could be thrown in "great droppes" like a fountain. The author says of ordinary means of garden watering: —

"The common Watring Pot with us hath a narrow Neck, a Big Belly, Somewhat large Bottome, and full of little holes with a proper hole forced in the head to take in the water; which filled full and the Thumbe laid on the hole to keep in the aire may in such wise be carried in handsome Manner."

Garden tools have changed but little since Tudor days; spade and rake were like ours to-day, so were dibble and mattock. Even grafting and pruning tools, shown in books of husbandry, were surprisingly like our own. Scythes were much heavier and clumsier. An old fellow is here shown sharpening in the ancient manner a scythe about three hundred years old.

The art of grafting, known since early days, formed an important part of the gardener's craft. Large share of ancient garden treatises is devoted to minute instructions therein. To this day in New England towns a good grafter is a local autocrat.

Beehives were once found in every garden; bee-skepes they were called when made of straw. Picturesque and homely were the old straw beehives, and still are they used in England; the old one shown in the chapter on sun-dials can scarcely be mated in America. They served as a conventional emblem of industry. They were made of welts or ropes of twisted straw, as were the heavy winnowing skepes once used for winnowing grain. In Maine, in a few out-of-the-way communities, ancient men still winnow grain with these skepes. I saw a man last autumn,

Summer-house at Ravensworth.

a giant in stature, standing in a dull light on the crown of a hill winnowing wheat in one of these

Sharpening the Old Dutch Scythe.

great skepes with an indescribably free and noble gesture. He was a classic, a relic of Homer's age, no longer a farmer, but a husbandman. Bees and honey were of much value in ancient days. Honey was the chief ingredient in many wholesome and pleasing drinks — mead, metheglin, bragget (or braket), morat, erboule — all very delightful in their ingredients, redolent of meadows and hedge-rows; thus Cowslip mead was made of Cowslip "pips," honey, Lemon juice, and "a handful of Sweetbrier." "Athol porridge," demure of name, was as potent as pleasing — potent as good honey, good cream, and good whiskey could make it.

Rows of typical Southern beehives are shown in the two succeeding illustrations. From their home by the side of a White Rose and under an old Sweet Apple tree these Waterford bees did not wish to swarm out in a hurry to find a new home. These beehives are not very ancient in shape, but when I see a row of them set thus under the trees, or in a hive-shelter, they seem to tell of olden days. The very bees flying in an out seem steady-going, respectable old fellows. Such hives have a cosy look, with rows of Hollyhocks behind them, and hundreds of spires of Larkspur for these old bees to bury their heads in.

The sadly picturesque old superstition of "telling the bees" of a death in a family and hanging a bit of black cloth on the hives as a mourning-weed still is observed in some country communities. Whittier's poem on the subject is wonderfully "countrified" in atmosphere, using the word chore-girl, so seldom heard even in familiar speech to-day and never found in verse elsewhere than in this rustic poem. I saw one summer in Narragansett, on Stony Lane, not far from the old Six-Principle Church, a row of beehives hung with strips of black cloth; the house mistress was dead — the friend of bird and beast and bee — who had reared the guardian of the garden told of on page 396 *et seq.*

A pretty and appropriate garden furnishing was the dove-cote. The possession of a dove-cote in England, and the rearing of pigeons, was free only to lords of the manor and noblemen. When the colonists came to America, many of them had never been permitted to keep pigeons. In Scotland persistent attemps at pigeon-raising by folks of humble station might be punished with death. The settlers must have revelled in the freedom of the new land, as well as in the plenty of pigeons, both wild and domestic. In old England the dove-cote was often built close to the kitchen door, that squab and pigeon might be near the hand of the cook. Dove-cotes in America were often simple boxes or houses raised on stout posts. Occasionally might be seen a fine brick dove-cote like the one still standing at Shirley-on-the-James, in Virginia, which is shaped without and within like several famous old dove-cotes in England, among them the one at Athelhampton Hall, Dorchester, England. The English dove-cote has within a revolving ladder hung from a central post while the Virginian squab catcher uses an ordinary ladder. The shelves for the birds to rest upon and the square

Beehives at Waterford, Virginia.

recesses for the nests made by the ingenious placing of the bricks are alike in both cotes.

A beautiful and fitting tenant of old formal gardens was the peacock, "with his aungelis federys bryghte." On large English estates peacocks were universally kept. A fine peacock, with full-spread

Beehives under the Trees.

tail, makes many a gay flower bed pale before his panoply of iridescence and color. The peahen is a demurely pretty creature. Peacocks are not altogether grateful to garden owners; on the old Narragansett farm whose garden is shown on page 35, they were always kept, and it was one of the prides and pleasures of formal hospitality to offer a roasted peacock to visitors. But, save when roasted, the vain creatures would not keep silence, and when they squawked the glory of their plumage was forgotten. They had an odious habit, too, of wandering off to distant groves on the farm, usually selecting the nights of bitterest cold, and roosting in some very high tree, in some very inaccessible spot. They could not be left in this ill-considered sleeping-place, else they would all freeze to death ; and words fail to tell the labor in lowering twilight and temperature of discovering their retreat, the dislodging, capturing, and imprisoning them.

In Narragansett there is a charming old farm garden, which I often visit to note and admire its old-time blossoms. This garden has a guardian, who haunts the garden walks as did the terrace peacock of old England ; no watch-dog ever was so faithful, and none half so acute. When I visit the garden I always ask " Where is Job ? " I am answered that he is in the field with the cattle. Sometimes this is true, but at other times Job has left the field and is attending to his assumed duties. As he is not encouraged, he has learned great slyness and dissimulation. Immovable, and in silence, Job is concealed behind a Syringa hedge or in a Lilac ambush, and as you stroll peacefully and unwittingly down the paths, sniffing the honeyed sweetness of the dense edging of Sweet Alyssum, all is as balmy as the blossoms. But stoop for an instant, to gather some leaves of Sweet Basil or Sweet Brier, or to collect a dozen seed-pods of that specially delicate Sweet Pea, and

Spring House at Johnson Homestead, Germantown, Pennsylvania.

lo ! the enemy is upon you, like a fierce whirlwind. He looks mild and demure enough in his kitchen yard retreat, whereto, upon piercing outcry for help, the farmer and his two sons have haled him, and

Dove-cote at Shirley-on-James.

where the camera has caught him. But far from meek is his aspect when you are dodging him around the great Tree Peony, or flying frantically before him down the side path to the garden gate. This fierce wild beast was once that mildest of creatures — a pet lamb; the constant companion of the farm-wife, as she weeded and watered her loved garden. Her husband says, "He seems to think folks are stealing her flowers, if they stop to look." The wife and mother of these three great men has gone from her garden forever; but a tenderness for all

The Peacock in His Pride.

that she loved makes them not only care for her flowers, but keeps this rampant guardian of the garden at the kitchen door, just as she kept him when he was a little lamb. I knew this New England farmer's wife, a noble woman, of infinite tenderness, strength, and endurance; a lover of trees and flowers and all living things, and I marvel not that they keep her memory green.

CHAPTER XIX

GARDEN BOUNDARIES

"A garden fair . . . with Wandis long and small
Railèd about, and so with treès set
Was all the place; and Hawthorne hedges knet,
That lyf was none walking there forbye
That might within scarce any wight espy."
— *Kings Qubair*, King James I of Scotland.

NE who reads what I have written in these pages of a garden enclosed, will scarcely doubt that to me every garden must have boundaries, definite and high. Three old farm boundaries were of necessity garden boundaries in early days — our stone walls, rail fences, and hedge-rows. The first two seem typically American; the third is an English hedge fashion. Throughout New England the great boulders were blasted to clear the rocky fields; and these, with the smaller loose stones, were gathered into vast stone walls. We still see these walls around fields and as the boundaries of kitchen gardens and farm flower gardens, and delightful walls they are, resourceful of beauty to the inventive gardener. I know one lovely garden in old Narragansett, on a farm which is now the country-seat of folk of great wealth, where the old stone walls are the pride of the place; and the carefully kept garden seems set in a beautiful frame of soft gray stones and flowering vines. These walls would be more beautiful still if our climate would let us have the wall gardens of old England, but

The Guardian of the Garden.

everything here becomes too dry in summer for wall gardens to flourish.

Rhode Island farmers for two centuries have cleared and sheltered the scanty soil of their state by blasting the ledges, and gathering the great stones of ledge and field into splendid stone walls. Their beauty is a gift to the farmer's descendants in reward for his hours of bitter and wearying toil. One of these fine stone walls, six feet in height, has stood secure and unbroken through a century of upheavals of winter frosts — which it was too broad and firmly built to heed. It stretches from the Post Road in old Narragansett, through field and meadow, and by the side of the oak grove, to the very edge of the bay. To the waterside one afternoon in June there strolled, a few years ago, a beautiful young girl and a somewhat conscious but determined young man. They seated themselves on the stone wall under the flickering shadow of a great Locust tree, then in full bloom. The air was sweet with the honeyed fragrance of the lovely pendent clusters of bloom, and bird and bee and butterfly hovered around,—it was paradise. The beauty and fitness of the scene so stimulated the young man's fancy to thoughts and words of love that he soon burst forth to his companion in an impassioned avowal of his desire to make her his wife. He had often pictured to himself that some time he would say to her these words, and he had seen also in his hopes the looks of tender affection with which she would reply. What was his amazement to behold that, instead of blushes and tender glances, his words of love were met by an apparently frenzied stare of horror and disgust, that seemed to pierce through him, as his beloved one sprung at one bound from her seat by his side on the high stone wall, and ran away at full speed, screaming out, "Oh, kill him! kill him!"

Now that was certainly more than disconcerting to the warmest of lovers, and with a half-formed dread that the suddenness of his proposal of love had turned her brain, he ran after her, albeit somewhat coolly, and soon learned the reason for her extraordinary behavior. Emulous of the tempting serpent of old, a great black snake, Mr. *Bascanion constrictor*, had said complaisantly to himself: "Now here are a fair young Adam and Eve who have entered uninvited my Garden of Eden, and the man fancies it is not good for him to be alone, but I will have a word to say about that. I will come to her with honied words." So he thrust himself up between the stones of the wall, and advanced persuasively upon them, behind the man's back. But a Yankee Eve of the year 1890 A.D. is not that simple creature, the Eve of the year —— B.C.; and even the Father of Evil would have to be great of guile to succeed in his wiles with her.

A farm servant was promptly despatched to watch for the ill-mannered and intrusive snake who — as is the fashion of a snake — had grown to be as big as a boa-constrictor after he vanished; and at the end of the week once more the heel of man had bruised the serpent's head, and the third party in this love episode lay dead in his six feet of ugliness, a silent witness to the truth of the story.

Throughout Narragansett, Locust trees have a fashion of fringing the stone walls with close young growth, and shading them with occasional taller trees. These form an ideal garden boundary. The stone walls also gather a beautiful growth of Clematis, Brier,

Terrace Wall at Van Cortlandt Manor.

wild Peas, and Grapes; but they form a clinging-place for that devil's brood, Poison Ivy, which is so persistent in growth and so difficult to exterminate.

The old worm fence was distinctly American; it had a zigzag series of chestnut rails, with stakes of twisted cedar saplings which were sometimes "chunked" by moss-covered boulders just peeping from the earth. This worm fence secured to the nature lover and to wild life a strip of land eight or ten feet wide, whereon plant, bird, beast, reptile, and insect flourished and reproduced. It has been, within a few years, a gardening fashion to preserve these old "Virginia" fences on country places of considerable elegance. Planted with Clematis, Honeysuckle, Trumpet vine, Wistaria, and the free-growing new Japanese Roses, they are wonderfully effective.

Rail Fence Corner.

On Long Island, east of Riverhead, where there are few stones to form stone walls, are curious and picturesque hedge-rows, which are a most interesting and characteristic feature of the landscape, and they are beautiful also, as I have seen them once or twice, at the end of an old garden. These hedgerows were thus formed: when a field was cleared, a row of young saplings of varied growth, chiefly

Topiary Work at Levens Hall.

Oak, Elder, and Ash, was left to form the hedge These young trees were cut and bent over parallel to the ground, and sometimes interlaced together with dry branches and vines. Each year these trees were lopped, and new sprouts and branches permitted to grow only in the line of the hedge. Soon a tangle of briers and wild vines overgrew and netted them all into a close, impenetrable, luxuriant mass. They were, to use Wordsworth's phrase, "scarcely hedge-rows, but lines of sportive woods run wild." In this close green wall birds build their nests, and in their shelter burrow wild hares, and there open Violets and other firstlings of the spring. The twisted tree trunks in these old hedges are sometimes three or four feet in diameter one way, and but a foot or more the other; they were a shiftless field-border, as they took up so much land, but they were sheep-proof. The custom of making a dividing line by a row of bent and polled trees still remains, even where the close, tangled hedge-row has disappeared with the flocks of sheep.

These hedge-rows were an English fashion seen in Hertfordshire and Suffolk. On commons and reclaimed land they took the place of the quickset hedges seen around richer farm lands. The bending and interlacing was called plashing; the polling, shrouding. English farmers and gardeners paid infinite attention to their hedges, both as a protection to their fields and as a means of firewood.

There is something very pleasant in the thought that these English gentlemen who settled eastern Long Island, the Gardiners, Sylvesters, Coxes, and others, retained on their farm lands in the new world the customs of their English homes, pleasanter still to know that their descendants for centuries kept up these homely farm fashions. The old hedge-rows on Long Island are an historical record, a landmark — long may they linger. On some of the finest estates on the island they have been carefully preserved, to form the lower boundary of a garden, where, laid out with a shaded, grassy walk dividing it from the flower beds, they form the loveliest of garden limits. Planted skilfully with great Art to look like great Nature, with edging of Elder and Wild Rose, with native vines and an occasional congenial garden ally, they are truly unique.

Yew was used for the most famous English hedges; and as neither Yew nor Holly thrive here — though both will grow — I fancy that is why we have ever had in comparison so few hedges, and have really no very ancient ones, though in old letters and account books we read of the planting of hedges on fine estates. George Washington tried it, so did Adams, and Jefferson, and Quincy. Osage Orange, Barberry, and Privet were in nurserymen's lists, but it has not been till within twenty or thirty years that Privet has become so popular. In Southern gardens, Cypress made close, good garden hedges; and Cedar hedges fifty or sixty years old are seen. Lilac hedges were unsatisfactory, save in isolated cases, as the one at Indian Hill. The Japan Quinces, and other of the Japanese shrubs, were tried in hedges in the mid-century, with doubtful success as hedges, though they form lovely rows of flowering shrubs. Snowballs and Snowberries, Flowering Currant, Altheas, and Locust, all have been used for hedge-planting, so we certainly have tried faithfully enough to have hedges in America. Locust hedges are most graceful, they cannot be clipped closely. I saw one lovely creation of Locust, set with an occasional Rose Acacia — and the Locust thus supported the brittle Aca-

Oval Pergola at Arlington.

cia. If it were successful, it would be, when in bloom, a dream of beauty. Hemlock hedges are ever fine, as are hemlock trees everywhere, but will not bear too close clipping. Other evergreens, among them the varied Spruces, have been set in hedges, but

French Homestead with old Stone Terrace, Kingston, Rhode Island.

have not proved satisfactory enough to be much used.

Buckthorn was a century ago much used for hedges and arches. When Josiah Quincy, President of Harvard College, was in Congress in 1809, he obtained from an English gardener, in Georgetown, Buckthorn plants for hedges in his Massachusetts home, which hedges were an object of great beauty for many years.

The traveller Kalm found Privet hedges in Pennsylvania in 1760. In Scotland Privet is called Primprint. Primet and Primprivet were other old names. Box was called Primpe. These were all derivative of prim, meaning precise. Our Privet hedges, new as they are, are of great beauty and satisfaction, and soon will rival the English Yew hedges.

I have never yet seen the garden in which there was not some boundary or line which could be filled to advantage by a hedge. In garden great or garden small, the hedge should ever have a place. Often a featureless garden, blooming well, yet somehow unattractive, has been completely transformed by the planting of hedges. They seem, too, to give such an orderly aspect to the garden. In level countries hedges are specially valuable. I cannot understand why some denounce clipped hedges and trees as against nature. A clipped hedge is just as natural as the cut grass of a lawn, and is closely akin to it. Others think hedges "too set"; to me their finality is their charm.

Hedges need to be well kept to be pleasing. Chaucer in his day in praising a "hegge" said that: —

> "Every branche and leaf must grow by mesure
> Pleine as a bord, of an height by and by."

In England, hedge-clipping has ever been a gardening art.

In the old English garden the topiarist was an important functionary. Besides his clipping shears

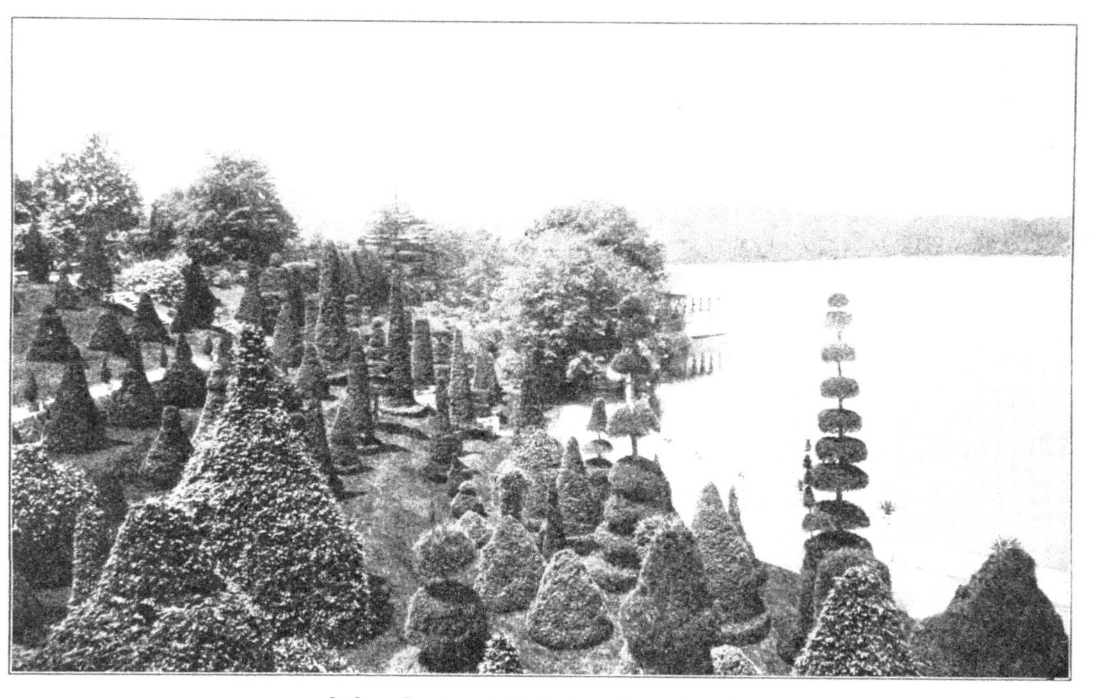

Italian Garden at Wellesley, Massachusetts.

he had to have what old-time cooks called *judgment* or *faculty*. In English gardens many specimens of topiary work still exist, maintained usually as relics of the past rather than as a modern notion of the beautiful. The old gardens at Levens Hall, page 404, contain some of the most remarkable examples.

In a few old gardens in America, especially in Southern towns, traces of the topiary work of early years can be seen; these overgrown, uncertain shapes have a curious influence, and the sentiment awakened is beautifully described by Gabriele d' Annunzio: —

"We walked among evergreens, among ancient Box trees, Laurels, Myrtles, whose wild old age had forgotten its early discipline. In a few places here and there was some trace of the symmetrical shapes carved once upon a time by the gardener's shears, and with a melancholy not unlike his who searches on old tombstones for the effigies of the forgotten dead, I noted carefully among the silent plants those traces of humanity not altogether obliterated."

The height of topiary art in America is reached in the lovely garden, often called the Italian garden, of Hollis H. Hunnewell, Esq., at Wellesley, Massachusetts. Vernon Lee tells in her charming essay on "Italian Gardens" of the beauty of gardens without flowers, and this garden of Mr. Hunnewell is an admirable example. Though the effect of the black and white of the pictured representations shown on these pages is perhaps somewhat sombre, there is nothing sad or sombre in the garden itself. The clear gleam of marble pavilions and balustrades, the formal rows of flower jars with their hundreds of Century plants, and the lovely light on the lovely lake, serve as a delightful contrast to the clear, clean lusty green of the clipped trees. This garden is a beautiful example of the art of the topiarist, not in its grotesque forms, but in the shapes liked by Lord Bacon, pyramids, columns, and "hedges in welts," carefully studied to be both stately and graceful. I first saw this garden thirty years ago; it was interesting then in its well thought-out plan, and in the perfection of every inch of its slow growth; but how much more beautiful now, when the garden's promise is fulfilled.

Steps in Italian Garden at Wellesley, Massachusetts.

The editor of *Country Life* says that the most notable attempt at modern topiary work in England is at Ascott, the seat of Mr. Leopold de Rothschild, but the examples there have not attained a growth at all approaching those at Wellesley. Mr. Hunnewell writes thus of his garden: —

"It was after a visit to Elvaston nearly fifty years ago that I conceived the idea of making a collection of trees for topiary work in imitation of what I had witnessed at that celebrated estate. As suitable trees for that purpose could not be obtained at the nurseries in this country, and as the English Yew is not reliable in our New England climate, I was obliged to make the best selection possible from such trees as had proved hardy here — the Pines, Spruces, Hemlocks, Junipers, Arbor-vitæ, Cedars, and Japanese Retinosporas. The trees were all very small, and for the first twenty years their growth was shortened twice annually, causing them to take a close and compact habit, comparing favorably in that respect with the Yew. Many of them are now more than forty feet in height and sixty feet in circumference, the Hemlocks especially proving highly successful."

This beautiful example of art in nature is ever open to visitors, and the number of such visitors is very large. It is, however, but one of the many beauties of the great estate, with its fine garden of Roses, its pavilion of splendid Rhododendrons and Azaleas, its uncommon and very successful rock garden, and its magnificent plantation of rare trees. There are also many rows of fine hedges and arches in various portions of the grounds, hedges of clipped Cedar and Hemlock, many of them twenty feet high, which compare well in condition, symmetry, and extent with the finest English hedges on the finest English estates.

Through the great number of formal gardens laid out within a few years in America, the topiary art has had a certain revival. In California, with

Topiary Work in California.

the lavish foliage, it may be seen in considerable perfection, though of scant beauty, as here shown.

Happy is the garden surrounded by a brick wall or with terrace wall of brick. How well every color looks by the side of old brick; even scarlet, bright pink, and rose-pink flowers, which seem impossible, do very well when held to the wall by clear green leaves. Flowering vines are perfect when trained on old soft-red brick enclosing walls; white-flowered vines are specially lovely thereon, Clematis, white Roses, and the rarely beautiful white Wistaria. How lovely is my Virgin's-bower when growing on brick;

Serpentine Brick Wall at University of Virginia, Charlottesville.

how Hollyhocks stand up beside it. Brick posts, too, are good in a fence, and, better still, in a pergola. A portion of the fine terrace wall at Van Cortlandt Manor is shown facing page 286. This wall was put in about fifty years ago; ere that there had been a grass bank, which is ever a trial in a garden; for it is hard to mow the grass on such a bank, and it never looks neat; it should be planted with some vine.

A very curious garden wall is the serpentine brick wall still standing at the University of Virginia, at Charlottesville. It is about seven feet high, and closes in the garden and green of the row of houses occupied by members of the faculty; originally it may have extended around the entire college grounds. I present a view from the street in order to show its contour distinctly; within the garden its outlines are obscured by vines and flowers. The first thought in the mind of the observer is that its reason for curving is that it could be built much more lightly, and hence more cheaply, than a straight wall; then it seems a possible idealization in brick of the old Virginia rail fence. But I do not look to domestic patterns and influences for its production; it is to me a good example of the old-time domination of French ideas which was so marked and so disquieting in America. In France, after the peace of 1762, the Marquis de Geradin was revolutionizing gardening. His own garden at Ermenonville and his description of it exercised important influence in England and America, as in France. Jefferson was the planner and architect of the University of Virginia; and it is stated that he built this serpentine wall. Whether he did or not, it is another example of French influences in architecture in the United States. This French school, above everything else, replaced straight lines with carefully curving and winding lines.

CHAPTER XX

A MOONLIGHT GARDEN

> "How sweetly smells the Honeysuckle
> In the hush'd night, as if the world were one
> Of utter peace and love and gentleness."
>
> — Walter Savage Landor.

ARDENS fanciful of name, a Saint's Garden, a Friendship Garden, have been planted and cherished. I plant a garden like none other; not an every-day garden, nor indeed a garden of any day, but a garden for "brave moonshine," a garden of twilight opening and midnight bloom, a garden of nocturnal blossoms, a garden of white blossoms, and the sweetest garden in the world. It is a garden of my dreams, but I know where it lies, and it now is smiling back at this very harvest moon.

The old house of Hon. Ben. Perley Poore — Indian Hill — at Newburyport, Massachusetts, has been for many years one of the loveliest of New England's homes. During his lifetime it had extraordinary charms, for on the noble hillside, where grew scattered in sunny fields and pastures every variety of native tree that would winter New England's snow and ice, there were vast herds of snow-white cows, and flocks of white sheep, and the splendid oxen were white. White pigeons circled in the air around ample dove-cotes, and the farm-yard poultry were all white; an enthusiastic chronicler recounts also white peacocks on the wall, but these are also denied.

On every side were old terraced walls covered with Roses and flowering vines, banked with shrubs, and standing in beds of old-time flowers running over with bloom; but behind the house, stretching up the lovely hillside, was The Garden, and when we entered it, lo! it was a White Garden with edgings of pure and seemly white Candytuft from the forcing beds, and flowers of Spring Snowflake and Star of Bethlehem and Jonquils; and there were white-flowered shrubs of spring, the earliest Spiræas and Deutzias; the doubled-flowered Cherries and Almonds and old favorites, such as Peter's Wreath, all white and wonderfully expressive of a simplicity, a purity, a closeness to nature.

I saw this lovely farmstead and radiant White Garden first in glowing sunlight, but far rarer must have been its charm in moonlight; though the white beasts (as English hinds call cattle) were sleeping in careful shelter; and the white dog, assured of their safety, was silent; and the white fowl were in coop and cote; and

> "Only the white sheep were sometimes seen
> To cross the strips of moon-blanch'd green."

But the White Garden, ah! then the garden truly lived; it was like lightest snow wreaths bathed in silvery moonshine, with every radiant flower adoring the moon with wide-open eyes, and pouring forth incense at her altar. And it was peopled with shadowy forms shaped of pearly mists and dews; and white night moths bore messages for them from flower to flower — this garden then was the garden of my dreams.

Thoreau complained to himself that he had not put duskiness enough into his words in his description of his evening walks. He longed to have the peculiar and classic severity of his sentences, the color of his style, tell his readers that his scene was laid at night without saying so in exact words. I, too, have not written as I wished, by moonlight; I can tell of moonlight in the garden, but I desire more; I want you to see and feel this moonlight garden, as did Emily Dickinson her garden by moonlight: —

"And still within the summer's night
A something so transporting bright
I clap my hands to see."

But perhaps I can no more gather it into words than I can bottle up the moonlight itself.

This lovely garden, varied in shape, and extending in many and diverse directions and corners, bears as its crown a magnificent double flower border over seven hundred feet long; with a broad straight path trimly edged with Box adown through its centre, and with a flower border twelve feet wide on either side. This was laid out and planted in 1833 by the parents of Major Poore, after extended travel in England,

and doubtless under the influences of the beautiful English flower gardens they had seen. Its length was originally broken halfway up the hill and crowned at the top of the hill by some formal parterres of careful design, but these now are removed. There are graceful arches across the path, one of Honeysuckle on the crown of the hill, from which you look out perhaps into Paradise — for Indian Hill in June is a very close neighbor to Paradise; it is difficult to define the boundaries between the two, and to me it would be hard to choose between them.

Standing in this arch on this fair hill, you can look down the long flower borders of color and perfume to the old house, lying in the heart of the trees and vines and flowers. To your left is the hill-sweep, bearing the splendid grove, an arboretum of great native trees, planted by Major Poore, and for which he received the prize awarded by his native state to the finest plantation of trees within its bounds. Turn from the house and garden, and look through this frame of vines formed by the arch upon this scene, — the loveliest to me of any on earth, — a fair New England summer landscape. Fields of rich corn and grain, broken at times with the gray granite boulders which show what centuries of grand and sturdy toil were given to make these fertile fields; ample orchards full of promise of fruit; placid lakes and mill-dams and narrow silvery rivers, with low-lying red brick mills embowered in trees; dark forests of sombre Pine and Cedar and Oak; narrow lanes and broad highways shaded with the

Chestnut Path in Garden at Indian Hill.

livelier green of Elm and Maple and Birch; gray farm-houses with vast barns; little towns of thrifty white houses clustered around slender church-spires which, set thickly over this sunny land, point everywhere to heaven, and tell, as if speaking, the story of New England's past, of her foundation on love of God, just as the fields and orchards and highways speak of thrift and honesty and hard labor; and the houses, such as this of Indian Hill, of kindly neighborliness and substantial comfort; and as this old garden speaks of a love of the beautiful, a refinement, an æsthetic and tender side of New England character which *we* know, but into which — as Mr. Underwood says in *Quabbin*, that fine study of New England life — "strangers and Kiplings cannot enter."

Seven hundred feet of double flower border, fourteen hundred feet of flower bed, twelve feet wide! "It do swallow no end of plants," says the gardener."

In spite of the banishing dictum of many artists in regard to white flowers in a garden, the presence of ample variety of white flowers is to me the greatest factor in producing harmony and beauty both by night and day. White seems to be as important a foil in some cases as green. It may sometimes be given to the garden in other ways than through flower blossoms, by white marble statues, vases, pedestals, seats.

We all like the approval of our own thoughts by men of genius; with my love of white flowers I had infinite gratification in these words of Walter Savage

Landor's, written from Florence in regard to a friend's garden: —

"I like white flowers better than any others; they resemble fair women. Lily, Tuberose, Orange, and the truly English Syringa are my heart's delight. I do not mean to say that they supplant the Rose and Violet in my affections, for these are our first loves, before we grew *too fond of considering*; and too fond of displaying our acquaintance with others of sounding titles."

In Japan, where flowers have rank, white flowers are the aristocrats. I deem them the aristocrats in the gardens of the Occident also.

Having been informed of Tennyson's dislike of white flowers, I have amused myself by trying to discover in his poems evidence of such aversion. I think one possibly might note an indifference to white blossoms; but strong color sense, his love of ample and rich color, would naturally make him name white infrequently. A pretty line in *Walking to the Mail* tells of a girl with "a skin as clean and white as Privet when it flowers"; and there were White Lilies and Roses and milk-white Acacias in Maud's garden.

In *The Last Tournament* the street-ways are depicted as hung with white samite, and "children sat in white," and the dames and damsels were all "white-robed in honor of the stainless child." A "swarthy one" cried out at last: —

"The snowdrop only, flowering thro' the year,
Would make the world as blank as wintertide.

Come ! — let us gladden their sad eyes
With all the kindlier colors of the field.
So dame and damsel glitter'd at the feast
Variously gay. . . .
So dame and damsel cast the simple white,
And glowing in all colors, the live grass,
Rose-campion, King-cup, Bluebell, Poppy, glanced
About the revels."

Foxgloves in Lower Garden at Indian Hill.

In the garden borders is a commonplace little plant, gray of foliage, with small, drooping, closed flowers of an indifferently dull tint, you would almost wonder at its presence among its gay garden fellows. Let us glance at it in the twilight, for it seems like the twilight, a soft, shaded gray ; but the flowers have already lifted their heads and opened their petals, and they now seem like the twilight clouds of palest pink and lilac. It is the Night-scented Stock, and lavishly through the still night it pours forth its ineffable fragrance. A single plant, thirty feet from an open window, will waft its perfume into the room. This white Stock was a favorite flower of Marie Antoinette, under its French name the Julienne. "Night Violets," is its appropriate German name. Hesperis! the name shows its habit. Dame's Rocket is our title for this cheerful old favorite of May, which shines in such snowy beauty at night, and throws forth such a compelling fragrance. It is rarely found in our gardens, but I have seen it growing wild by the roadside in secluded spots; not in ample sheets of growth like Bouncing Bet, which we at first glance thought it was; it is a shyer stray, blossoming earlier than comely Betsey.

The old-fashioned single, or slightly double, country Pink, known as Snow Pink or Star Pink, was often used as an edging for small borders, and its bluish green, almost gray, foliage was quaint in effect and beautiful in the moonlight. When seen at night, the reason for the folk-name is evident. Last summer, on a heavily clouded night in June, in a cottage garden at West Hampton, borders of this Snow Pink shone out of the darkness with a phosphorescent light, like hoar-frost, on every grassy leaf; while the hundreds of pale pink blossoms seemed softly shining stars. It was a curious effect, almost wintry, even in midsummer. The scent was wafted down the garden path, and along the country road, like a concentrated essence, rather than a fleeting breath of flowers. One of these cottage borders is shown on page 292, and I have named it from these lines from *The Garden that I Love*: —

> "A running ribbon of perfumed snow
> Which the sun is melting rapidly."

At sundown the beautiful white Day Lily opens and gives forth all night an overwhelming sweetness; I have never seen night moths visiting it, though I know they must, since a few seed capsules always form. In the border stand —

> "Clumps of sunny Phlox
> That shine at dusk, and grow more deeply sweet."

These, with white Petunias, are almost unbearably cloying in their heavy odor. It is a curious fact that some of these night-scented flowers are positively offensive in the daytime; try your *Nicotiana affinis* next midday — it outpours honeyed sweetness at night, but you will be glad it withholds its perfume by day. The plants of Nicotiana were first introduced to England for their beauty, sweet scent, and medicinal qualities, not to furnish smoke. Parkinson in 1629 writes of Tobacco, "With us it is cherished for medicinal qualities as for the beauty of its flowers," and Gerarde, in 1633, after telling of the beauty, etc., says that the dried leaves are "taken in a pipe, set on fire, the smoke suckt into the stomach, and thrust forth at the noshtrils."

Snake-root, sometimes called Black Cohosh (*Cimicifuga racemosa*), is one of the most stately wild flowers, and a noble addition to the garden. A picture of a single plant gives little impression of its dignity of habit, its wonderfully decorative growth; but the succession of pure white spires, standing up several feet high at the edge of a swampy field, or in a garden, partake of that compelling charm which comes from tall trees of slender growth, from repetition and association, such as pine trees, rows of

Dame's Rocket.

bayonets, the gathered masts of a harbor, from stalks of corn in a field, from rows of Foxglove — from all " serried ranks." I must not conceal the fact of its horrible odor, which might exile it from a small garden.

Among my beloved white flowers, a favorite among those who are all favorites, is the white Columbine. Some are double, but the common single white Columbines picture far better the derivation of their name; they are like white doves, they seem almost an emblematic flower. William Morris says: —

> "Be very shy of double flowers; choose the old Columbine where the clustering doves are unmistakable and distinct, not the double one, where they run into mere tatters. Don't be swindled out of that wonder of beauty, a single Snowdrop; there is no gain and plenty of loss in the double one."

There are some extremists, such as Dr. Forbes Watson, who condemn all double flowers. One thing in the favor of double blooms is that their perfume is increased with their petals. Double Violets, Roses, and Pinks seem as natural now as single flowers of their kinds. I confess a distinct aversion to the thought of a double Lilac. I have never seen one, though the Ranoncule, said to be very fine, costs but forty cents a plant, and hence must be much grown.

There is a curious influence of flower-color which I can only explain by giving an example. We think of Iris, Gladiolus, Lupine, and even Foxglove and Poppy as flowers of a warm and vivid color; so where we see them a pure white, they have a distinct and compelling effect on us, pleasing, but a little eerie; not a surprise, for we have always known the white varieties, yet not exactly what we are wonted to. This has nothing of the grotesque, as is produced by the albino element in the animal world; it is simply a trifle mysterious. White Pansies and

White Violets possess this quality to a marked degree. I always look and look again at growing White Violets. A friend says: "Do you think

Snake-root.

they will speak to you?" for I turn to them with such an expectancy of something.

The "everlasting" white Pea is a most satisfactory plant by day or night. Hedges covered with it are a pure delight. Do not fear to plant it with liberal hand. Be very liberal, too, in your garden of white Foxgloves. Even if the garden be small, there is room for many graceful spires of the lovely bells to shine out everywhere, piercing up through green foliage and colored blooms of other plants. They are not only beautiful, but they are flowers of sentiment and association, endeared to childhood, visited of bees, among the best beloved of old-time favorites. They consort well with nearly every other flower, and certainly with every other color, and they seem to clarify many a crudely or dingily tinted flower; they are as admirable foils as they are principals in the garden scheme. In England, where they readily grow wild, they are often planted at the edge of a wood, or to form vistas in a copse. I doubt whether they would thrive here thus planted, but they are admirable when set in occasional groups to show in pure whiteness against a hedge. I say in occasional groups, for the Foxglove should never be planted in exact rows. The White Iris, the Iris of the Florentine Orris-root, is one of the noblest plants of the whole world; its pure petals are truly hyaline like snow-ice, like translucent white glass; and the indescribably beautiful drooping lines of the flowers are such a contrast with the defiant erectness of the fresh green leaves. Small wonder that it was a sacred flower of the

Greeks. It was called by the French *la flambe blanche*, a beautiful poetic title — the White Torch of the Garden.

A flower of mystery, of wonderment to children, was the Evening Primrose; I knew the garden variety only with intimacy. Possibly the wild flower had similar charms and was equally weird in the gloaming, but it grew by country roadsides, and I was never outside our garden limits after nightfall, so I know not its evening habits. We had in our garden a variety known as the California Evening Primrose — a giant flower as tall as our heads. My mother saw its pale yellow stars shining in the early evening in a cottage garden on Cape Ann, and was there given, out of the darkness, by a fellow flower lover, the seeds which have afforded to us every year since so much sentiment and pleasure. The most exquisite description of the Evening Primrose is given by Margaret Deland in her *Old Garden*: —

"There the primrose stands, that as the night
Begins to gather, and the dews to fall,
Flings wide to circling moths her twisted buds,
That shine like yellow moons with pale cold glow,
And all the air her heavy fragrance floods,
And gives largess to any winds that blow.
Here in warm darkness of a night in June,
. . . children came
To watch the primrose blow. Silently they stood
Hand clasped in hand, in breathless hush around,
And saw her slyly doff her soft green hood
And blossom — with a silken burst of sound."

The Title-page of Parkinson's *Paradisi in Solis*, etc.

The wild Primrose opens slowly, hesitatingly, it trembles open, but the garden Primrose flares open.

The Evening Primrose is usually classed with sweet-scented flowers, but that exact observer, E. V. B., tells of its "repulsive smell. At night if the stem be shaken, or if the flower-cup trembles at the touch of a moth as it alights, out pours the dreadful odor." I do not know that any other garden flower opens with a distinct sound. Owen Meredith's poem, *The Aloe*, tells that the Aloe opened with such a loud explosive report that the rooks shrieked and folks ran out of the house to learn whence came the sound.

The tall columns of the Yucca or Adam's Needle stood like shafts of marble against the hedge trees of the Indian Hill garden. Their beautiful blooms are a miniature of those of the great Century Plant. In the daytime the Yucca's blossoms hang in scentless, greenish white bells, but at night these bells lift up their heads and expand with great stars of light and odor — a glorious plant. Around their spire of luminous bells circle pale night moths, lured by the rich fragrance. Even by moonlight we can see the little white detached fibres at the edge of the leaves, which we are told the Mexican women used as thread to sew with. And we children used to pull off the strong fibres and put them in a needle and sew with them too.

When I see those Yuccas in bloom I fully believe that they are the grandest flowers of our gardens; but happily, I have a short garden memory, so I mourn not the Yucca when I see the *Anemone japonica* or any other noble white garden child.

Yucca, like White Marble against the Evergreens.

Here at the end of the garden walk is an arbor dark with the shadow of great leaves, such as Gerarde calls "leaves round and big like to a buckler."

But out of that shadowed background of leaf on leaf shine hundreds of pure, pale stars of sweetness and light, — a true flower of the night in fragrance, beauty, and name, — the Moon-vine. It is a flower of sentiment, full of suggestion.

Did you ever see a ghost in a garden? I do so wish I could. If I had the placing of ghosts, I would not make them mope round in stuffy old bedrooms and garrets; but would place one here in this arbor in my Moonlight Garden. But if I did, I have no doubt she would take up a hoe or a watering-pot, and proceed to do some very unghostlike deed — perhaps, grub up weeds. Longfellow had a ghost in his garden (page 142). He must have mourned when he found it was only a clothes-line and a long night-gown.

It was the favorite tale of a Swedish old lady who lived to be ninety-six years old, of a discovery of her youth, in the year 1762, of strange flashes of light which sparkled out of the flowers of the Nasturtium one sultry night. I suppose the average young woman of the average education of the day and her country might not have heeded or told of this, but she was the daughter of Linnæus, the great botanist, and had not the everyday education.

Then great Goethe saw and wrote of similar flashes of light around Oriental Poppies; and soon other folk saw them also — naturalists and everyday folk. Usually yellow flowers were found to display this light — Marigolds, orange Lilies, and Sunflowers. Then the daughter of Linnæus reported another curious discovery; she certainly turned her nocturnal rambles in her garden to good account. She averred she had set fire to a certain gas which formed and hung around the Fraxinella, and that the ignition did not injure the plant. This assertion was met with open scoffing and disbelief, which has never wholly ceased; yet the popular name of Gas Plant indicates a widespread confidence in this quality of the Fraxinella and it is easily proved true.

Another New England name for the Fraxinella, given me from the owner of the herb-garden at Elmhurst, is "Spitfire Plant," because the seed-pods sizzle so when a lighted match is applied to them.

The Fraxinella is a sturdy, hardy flower. There are some aged plants in old New England gardens; I know one which has outlived the man who planted it, his son, grandson, and great-grandson. The Fraxinella bears a tall stem with Larkspur-like flowers of white or a curious dark pink, and shining Ash-like leaves, whence its name, the little Ash. It is one of the finest plants of the old-fashioned garden; fine in bloom, fine in habit of growth, and it even has decorative seed vessels. It is as ready of scent as anything in the garden; if you but brush against leaf, stem, flower, or seed, as you walk down the garden path, it gives forth a penetrating perfume, that you think at first is like Lemon, then like Anise, then like Lavender; until you finally decide it is like nothing save Fraxinella. As with the blossoms of the Calycanthus shrub, you can never mistake the perfume, when once you know it, for anything else. It is a scent of distinction. Through this individuality it is, therefore, full of associations, and correspondingly beloved.

Fraxinella.

CHAPTER XXI

FLOWERS OF MYSTERY

"Let thy upsoaring vision range at large
This garden through: for so by ray divine
Kindled, thy ken a magic flight shall mount."
— CARY'S Translation of Dante.

OGIES and fairies, a sense of eeriness, came to every garden-bred child of any imagination in connection with certain flowers. These flowers seemed to be regarded thus through no special rule or reason. With some there may have been slight associations with fairy lore, or medicinal usage, or a hint of meretriciousness. Sometimes the child hardly formulated his thought of the flower, yet the dread or dislike or curiosity existed. My own notions were absolutely baseless, and usually absurd. I doubt if we communicated these fancies to each other save in a few cases, as of the Monk's-hood, when we had been warned that the flower was poisonous.

I have read with much interest Dr. Forbes Watson's account of plants that filled his childish mind with mysterious awe and wonder; among them were the Spurge, Henbane, Rue, Dogtooth Violet, Nigella, and pink Marsh Mallow. The latter has ever been to me one of the most cheerful of blossoms. I did not know it in my earliest childhood, and never saw it in gardens till recent years. It is too close a cousin of the Hollyhock ever to seem to me aught but a happy flower. Henbane and Rue I did not know, but I share his feeling toward the others, though I could not carry it to the extent of fancying these the plants which a young man gathered, distilled, and gave to his betrothed as a poison.

There has ever been much uncanny suggestion in the Cypress Spurge. I never should have picked it had I found it in trim gardens; but I saw it only in forlorn and neglected spots. Perhaps its sombre tinge may come now from association, since it is often seen in country graveyards; and I heard a country woman once call it "Graveyard Ground Pine." But this association was not what influenced my childhood, for I never went then to graveyards.

In driving along our New England roads I am ever reminded of Parkinson's dictum that "Spurge once planted will hardly be got rid out again." For by every decaying old house, in every deserted garden, and by the roadside where houses may have been, grows and spreads this Cypress Spurge. I know a large orchard in Narragansett from which grass has wholly vanished; it has been crowded out by the ugly little plant, which has even invaded the adjoining woods.

I wonder why every one in colonial days planted

it, for it is said to be poisonous in its contact to some folks, and virulently poisonous to eat — though I am sure no one ever wanted to eat it. The colonists even brought it over from England, when we had here such lovely native plants. It seldom flowers. Old New England names for it are Love-in-a-huddle and Seven Sisters; not over significant, but of interest, as folk-names always are.

I join with Dr. Forbes Watson in finding the Nigella uncanny. It has a half-spidery look, that seems ungracious in a flower. Its names are curious: Love-in-a-mist, Love-in-a-puzzle, Love-in-a-tangle, Puzzle-love, Devil-in-a-bush, Katherine-flowers — another of the many allusions to St. Katherine and her wheel; and the persistent styles do resemble the spokes of a wheel. A name given it in a cottage garden in Wayland was Blue Spider-flower, which seems more suited than that of Spiderwort for the Tradescantia. Spiderwort, like all "three-cornered" flowers, is a flower of mystery; and so little cared for to-day that it is almost extinct in our gardens, save where it persists in out-of-the-way spots. A splendid clump of it is here shown, which grows still in the Worcester garden I so loved in my childhood. In this plant the old imagined tracings of spider's legs in the leaves can scarce be seen. With the fanciful notion of "like curing like" ever found in old medical recipes, Gerarde says, vaguely, the leaves are good for "the Bite of that Great Spider," a creature also of mystery.

Perhaps if the clear blue flowers kept open

throughout the day, the Spiderwort would be more tolerated, for this picture certainly has a Japanesque appearance, and what we must acknowledge was far more characteristic of old-time flowers than of many new ones, a wonderful individuality; there was no sameness of outline. I could draw the outline of a dozen blossoms of our modern gardens, and you could not in a careless glance distinguish one from the other: Cosmos, *Anemone japonica*, single Dahlias, and Sunflowers, Gaillardia, Gazanias, all such simple Rose forms.

Love-in-a-mist.

There was a quaint and mysterious annual in ancient gardens, called Shell flower, or Molucca Balm, which is not found now even on seedsmen's special lists of old-fashioned plants. The flower was white, pink-tipped, and set in a cup-shaped calyx an inch long, which was bigger than the flower itself. The plant stood two or three feet high, and the sweet-scented flowers were in whorls of five or six on a stem. It is a good example of my assertion that the old flowers had queerer shapes than modern ones, and were made of queer materials; the calyx of this Shell flower is of such singular quality and fibre.

The Dog-tooth Violet always had to me a sickly look, but its leaves give it its special offensiveness; all spotted leaves, or flower petals which showed the slightest resemblance to the markings of a snake or lizard, always filled me with dislike. Among them I included Lungwort (Pulmonaria), a flower which seems suddenly to have disappeared from many gardens, even old-fashioned ones, just as it has disappeared from medicine. Not a gardener could be found in our public parks in New York who had ever seen it, or knew it, though there is in Prospect Park a well-filled and noteworthy "Old-fashioned Garden." Let me add, in passing, that nothing in the entire park system—greenhouses, water gardens, Italian gardens—affords such delight to the public as this old-fashioned garden.

The changing blue and pink flowers of the Lungwort, somewhat characteristic of its family, are curious also. This plant was also known by the singular name of Joseph-and-Mary; the pink flowers being the emblem of Joseph; the blue of the Blessed Virgin Mary. Lady's-tears was an allied name, from a legend that the Virgin Mary's tears fell on the leaves, causing the white spots to grow in them, and that one of her blue eyes became red from excessive weeping. It was held to be unlucky even to destroy the plant. Soldier-and-his-wife also had reference to the red and blue tints of the flower.

A cousin of the Lungwort, our native *Mertensia virginica*, has in the young plant an equally singular leafage; every ordinary process of leaf progress is reversed: the young shoots are not a tender green, but are almost black, and change gradually in leaf, stem, and flower calyx to an odd light green in which the dark color lingers in veins and spots until the plant is in its full flower of tender blue, lilac, and pink. "Blue and pink ladies" we used to call the blossoms when we hung them on pins for a fairy dance.

The Alstrœmeria is another spotted flower of the old borders, curious in its funnel-shaped blooms, edged and lined with tiny brown and green spots. It is more grotesque than beautiful, but was beloved in a day that deemed the Tiger Lily the most beautiful of all lilies.

The aversion I feel for spotted leaves does not extend to striped ones, though I care little for variegated or striped foliage in a garden. I like the striped white and green leaves of one variety of our garden Iris, and of our common Sweet Flag (Calamus), which are decorative to a most satisfactory degree. The firm ribbon leaves of the striped Sweet Flag never turn brown in the driest summer, and grow very tall; a tub of it kept well watered is a thing of surprising beauty, and the plants are very handsome in the rock garden. I wonder what the

Spiderwort.

bees seek in the leaves! they throng its green and white blades in May, finding something, I am sure, besides the delightful scent; though I do not note that they pierce the veins of the plant for the sap, as I have known them to do along the large veins of certain palm leaves. I have seen bees often act as though they were sniffing a flower with appreciation, not gathering honey. The only endeared striped leaf was that of the Striped Grass — Gardener's Garters we called it. Clumps of it growing at Van Cortlandt Manor are here shown. We children used to run to the great plants of Striped Grass at the end of the garden as to a toy ribbon shop. The long blades of Grass looked like some antique gauze ribbons. They were very modish for dolls' wear, very useful to shape pin-a-sights, those very useful things, and very pretty to tie up posies. Under favorable circumstances this garden child might become a garden pest, a spreading weed. I never saw a more curious garden stray than an entire dooryard and farm garden — certainly two acres in extent, covered with Striped Grass, save where a few persistent Tiger Lilies pierced through the striped leaves. Even among the deserted hearthstones and tumble-down chimneys the striped leaves ran up among the roofless walls.

Let me state here that the suggestion of mystery in a flower did not always make me dislike it; sometimes it added a charm. The Periwinkle — Ground Myrtle, we used to call it — was one of the most mysterious and elusive flowers I knew, and other children thus regarded it; but I had a deep affection for its lovely blue stars and clean, glossy leaves, a special love, since it was the first flower I saw

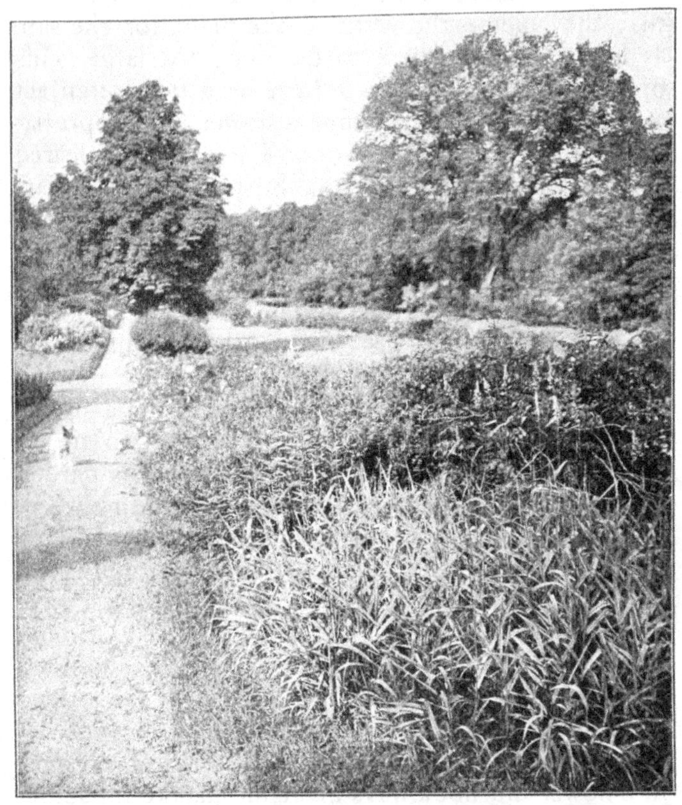

Gardener's Garters, at Van Cortlandt Manor.

blooming out of doors after a severe illness, and it seemed to welcome me back to life.

The name is from the French Pervenche, which suffers sadly by being changed into the clumsy Periwinkle. Everywhere it is a flower of mystery; it is the "Violette des Sorciers" of the French. Sadder is its Tuscan name, "Flower of death," for it is used there as garlands at the burial of children; and is often planted on graves, just as it is here. A far happier folk-name was Joy-of-the-ground, and to my mind better suited to the cheerful, healthy little plant.

An ancient medical manuscript gives this description of the Periwinkle, which for directness and lucidity can scarcely be excelled:—

> "Parwyke is an erbe grene of colour,
> In tyme of May he bereth blue flour.
> Ye lef is thicke, schinede and styf,
> As is ye grene jwy lefe.
> Vnder brod and uerhand round,
> Men call it ye joy of grownde."

On the list of the Boston seedsman (given on page 33 *et seq.*) is Venus'-navelwort. I lingered this summer by an ancient front yard in Marblehead, and in the shade of the low-lying gray-shingled house I saw a refined plant with which I was wholly unacquainted, lying like a little dun cloud on the border, a pleasing plant with cinereous foliage, in color like the silvery gray of the house, shaded with a bluer tint and bearing a dainty milk-white bloom. This modest flower had that power of catching the attention in spite of the high and striking colors of its neighbors, such as a simple gown of gray and white, if of graceful cut and shape, will have among gay-colored silk attire — the charm of Quaker garb, even though its shape be ugly. You know how ready is the owner of such a garden to talk of her favorites, and soon I was told that this plant was "Navy-work." I accepted this name in this old maritime town as possibly a local folk-name, yet I was puzzled by a haunting memory of having heard some similar title. A later search in a botany revealed the original, Venus'-navelwort.

I deem it right to state in this connection that any such corruption of the old name of a flower is very unusual in Massachusetts, where the English tongue is spoken by all of Massachusetts descent in much purity of pronunciation.

There is no doubt that all the flowers of the old garden were far more suggestive, more full of meaning, than those given to us by modern florists. This does not come wholly from association, as many fancy, but from an inherent quality of the flower itself. I never saw Honeywort (Cerinthe) till five years ago, and then it was not in an old-fashioned garden; but the moment I beheld the graceful, drooping flowers in the flower bed, the yellow and purple-toothed corolla caught my eye, as it caught my fancy; it seemed to mean something. I was not surprised to learn that it was an ancient favorite of colonial days. The leaves of Honeywort are often lightly spotted, which may be one of its elements of mystery. Honeywort is seldom seen even in our oldest gardens; but it is a beautiful flower and a most hardy annual, and deserves to be reintroduced.

Garden Walk at The Manse, Deerfield, Massachusetts.

A great favorite in the old garden was the splendid scarlet Lychnis, to which in New England is given the name of London Pride. There are two old varieties: one has four petals with squared ends, and is called, from the shape of the expanded flower, the Maltese Cross; the other, called Scarlet Lightning, is shown on a succeeding page; it has five deeply-nicked petals. It is a flower of midsummer eve and magic power, and I think it must have some connection with the Crusaders, being called by Gerarde Floure of Jerusalem, and Flower of Candy. The five-petalled form is rarely seen; in one old family I know it is so cherished, and deemed so magic a home-maker, that every bride who has gone from that home for over a hundred years has borne away a plant of that London Pride; it has really become a Family Pride.

Another plant of mysterious suggestion was the common Plantain. This was not an unaided instinct of my childhood, but came to me through an explanation of the lines in the chapter, "The White Man's Foot," in *Hiawatha*: —

> "Whereso'er they tread, beneath them
> Springs a flower unknown among us;
> Springs the White Man's Foot in blossom."

After my father showed me the Plantain as the "White Man's Foot," I ever regarded it with a sense of its unusual power; and I used often to wonder, when I found it growing in the grass, who had stepped there. I have permanently associated with the Plantain or Waybred a curious and distasteful trick of my memory. We recall our American humorist's lament over the haunting lines from the car-conductor's orders, which filled his brain and ears from the moment he read them, wholly by chance, and which he tried vainly to forget. A similar obsession filled me when I read the spirited apostrophe to the Plantain or Waybred, in Cockayne's translation of Ælfric's *Lacunga*, a book of leechcraft of the eleventh century: —

"And thou Waybroad,
Mother of worts,
Over thee carts creaked,
Over thee Queens rode,
Over thee brides bridalled,
Over thee bulls breathed,
All these thou withstoodst,
Venom and vile things,
And all the loathly things,
That through the land rove."

I could not thrust them out of my mind; worse still, I kept manufacturing for the poem scores of lines of similar metre. I never shall forget the Plantain, it won't let me forget it.

The Orpine was a flower linked with tradition and mystery in England, there were scores of fanciful notions connected with it. It has grown to be a spreading weed in some parts of New England, but it has lost both its mystery and its flowers. The only bed of flowering Orpine I ever saw in America was in the millyard of Miller Rose at Kettle Hole — and a really lovely expanse of bloom it was, broken only by old worn millstones which formed the doorsteps. He told with pride that his grandmother planted it, and "it was the flowering variety that no one else had in Rhode Island, not even in greenhouses in Newport." Miller Rose ground corn meal and flour with ancient millstones, and infinitely better were his grindings than "store meal." He could tell you, with prolonged detail, of the new-fangled roller he bought and used one week, and not a decent Johnny-cake could be made from the meal, and it shamed him. So he threw away all the meal he hadn't sold; and then the new machinery was pulled out and the millstones replaced, "to await the Lord's coming," he added, being a Second Adventist — or by his own title a "Christadelphian and an Old Bachelor." He was a famous preacher, having a pulpit built of heavy stones, in the woods near his mill. A little trying it was to hear the outpourings of his long sermons on summer afternoons, while you waited for him to come down from his pulpit and his prophesyings to give you your bag of meal. A tithing of time he gave each day to the

London Pride.

Lord, two hours and a half of preaching — and doubtless far more than a tithe of his income to the poor. In sentimental association with his name, he had a few straggling Roses around his millyard — all old-time varieties; and, with Orpine and Sweet-brier, he could gather a very pretty posy for all who came to Kettle Hole.

We constantly read of Fritillaries in the river fields sung of Matthew Arnold. In a charming book of English country life, *Idlehurst*, I read how closely the flower is still associated with Oxford life, recalling ever the Iffley and Kensington meadows to all Oxford men. The author tells that "quite unlikely sorts of men used to pick bunches of the flowers, and we would come up the towpath with our spoils." Fritillaries grew in my mother's garden; I cannot now recall another garden in America where I have ever seen them in bloom. They certainly are not common. On a succeeding page are shown the blossoms of the white Fritillary my mother planted and loved. Can you not believe that we love them still? They have spread but little, neither have they dwindled nor died. Each year they seem to us the very same blossoms she loved.

Our cyclopædias of gardening tell us that the Fritillaries spread freely; but E. V. B. writes of them in her exquisite English: "Slow in growth as the Fritillaries are, they are ever sure. When they once take root, there they stay forever, with a constancy unknown in our human world. They may be trusted, however late their coming. In the fresh vigor of its youth was there ever seen any other flower planned so exquisitely, fashioned so slenderly! The pink symmetry of Kalmia perhaps comes nearest this perfection, with the delicately curved and rounded angles of its bloom."

In no garden, no matter how modern, could the Fritillaries ever look to me aught but antique and classic. They are as essentially of the past, even to the careless eye, as an antique lamp or brazier. Quaint, too, is the fabric of their coats, like some old silken stuff of paduasoy or sarsenet. All are checkered, as their name indicates. Even the white flowers bear little birthmarks of checkered lines. They were among the famous dancers in my mother's garden, and I can tell you that a country dance of Fritillaries in plaided kirtles and green caps is a lively sight. Another name for this queer little flower is Guinea-hen Flower. Gerarde, with his felicity of description, says:—

> "One square is of a greenish-yellow colour, the other purple, keeping the same order as well on the back side of the flower as on the inside; although they are blackish in one square, and of a violet colour in another: in so much that every leafe (of the flower) seemeth to be the feather of a Ginnie hen, whereof it took its name."

A strong personal trait of the Fritillaries (for I may so speak of flowers I love) is their air of mystery. They mean something I cannot fathom; they look it, but cannot tell it. Fritillaries were a flower of significance even in Elizabethan days. They were made into little buttonhole posies, and, as Parkinson says, "worn abroad by curious lovers of these delights." In California grow wild a dozen varieties; the best known of these is recurved, but it does not droop, and is to all outward glance an Anemone, and has lost in that new world much the mystery of the old herbalist's "Checker Lily," save the checkers; these always are visible.

The Cyclamen and Dodecatheon lay their ears back like a vicious horse. Both have an eerie aspect,

White Fritillaria.

as if turned upside down, as has also the Nightshade. I knew a little child, a flower lover from babyhood, who feared to touch the Cyclamen, and even cried if any attempt was made to have her touch the flower. When older, she said that she had feared the flower would sting her.

I have often a sense of mysterious meaning in a vine, it seems so plainly to reach out to attract your attention. I recall once being seated on the door-step of a deserted farm-house, musing a little over the sad thought of this lost home, when suddenly some one tapped me on the cheek — I suppose I ought to say some thing, though it seemed a human touch. It was a spray of Matrimony vine, twenty feet long or more, that had reached around a corner, and helped by a breeze, had appealed to me for sympathy and companionship. I answered by following it around the corner. It had been trained up to a little shelf-like ledge or roof, over what had been a pantry window, and hung in long lines of heavy shade. It said to me: "Here once lived a flower-loving woman and a man who cared for her comfort and pleasure. She planted me when she, and the man, and the house were young, and he made the window shelter, and trained me over it, to make cool and green the window where she worked. I was the symbol of their happy married love. See! there they lie, under the gray stone beneath those cedars. Their children all are far away, but every year I grow fresh and green, though I find it lonely here now." To me, the Matrimony vine is ever a plant of interest, and it may be very beautiful, if cared for. On page 186 is shown the lovely growth on the porch at Van Cortlandt Manor.

With a sentiment of wonder and inquiry, not unmixed with mystery, do we regard many flowers, which are described in our botanies as Garden Escapes. This Matrimony vine is one of the many creeping, climbing things that have wandered away from houses. Honeysuckles and Trumpet-vines are far travellers. I saw once in a remote and wild spot a great boulder surrounded with bushes and all were covered with the old Coral or Trumpet Honeysuckle; it had such a familiar air, and yet seemed to have gained a certain knowingness by its travels.

This element of mystery does not extend to the flowers which I am told once were in trim gardens, but which I have never seen there, such as Ox-eye Daisies, Scotch Thistles, Chamomile, Tansy, Bergamot, Yarrow, and all of the Mint family; they are to me truly wild. But when I find flowers still cherished in our gardens, growing also in some wild spot, I regard them with wonder. A great expanse of Coreopsis, a field of Grape Hyacinth or Star of Bethlehem, roadsides of Coronilla or Moneywort, rows of red Day Lily and Tiger Lily, patches of Sunflowers or Jerusalem Artichokes, all are matters of thought; we long to trace their wanderings, to have them tell whence and how they came. Bouncing Bet is too cheerful and rollicking a wanderer to awaken sentiment. How gladly has she been welcomed to our fields and roadsides. I could not willingly spare her in our country drives, even to become again a cherished garden dweller. She rivals the Succory in beautifying arid dust heaps and barren railroad cuts, with her tender opalescent pink tints. How wholesome and hearty her growth, how pleasant her fragrance. We can never see her too often, nor ever stigmatize her, as have been so many of our garden escapes, as "Now a dreaded weed."

One of the weirdest of all flowers to me is the

Butter-and-eggs, the Toad-flax, which was once a garden child, but has run away from gardens to wander in every field in the land. I haven't the slightest reason for this regard of Butter-and-eggs, and I believe it is peculiar to myself, just as is Dr. Forbes Watson's regard of the Marshmallow to him. I

Bouncing Bet.

have no uncanny or sad associations with it, and I never heard anything "queer" about it. Thirty years ago, in a locality I knew well in central Massachusetts, Butter-and-eggs was far from common; I even remember the first time I saw it and was told its quaint name; now it grows there and everywhere; it is a persistent weed. John Burroughs calls it "the hateful Toad-flax," and old Manasseh Cutler, in a curious mixture of compliment and slur, "a common, handsome, tedious weed." It travels above ground and below ground, and in some soils will run out the grass. It knows how to allure the bumblebee, however, and has honey in its heart. I think it a lovely flower, though it is queer; and it is a delight to the scientific botanist, in the delicate perfection of its methods and means of fertilization.

The greatest beauty of this flower is in late autumn, when it springs up densely in shaven fields. I have seen, during the last week in October, fields entirely filled with its exquisite sulphur-yellow tint, one of the most delicate colors in nature; a yellow that is luminous at night, and is rivalled only by the pale yellow translucent leaves of the Moosewood in late autumn, which make such a strange pallid light in old forests in the North — a light which dominates over every other autumn tint, though the trees which bear them are so spindling and low, and little noted save in early spring in their rare pinkness, and in this their autumn etherealization. And the Moosewood shares the mystery of the Butter-and-eggs as well as its color. I should be afraid to drive or walk alone in a wood road, when the Moosewood leaves were turning yellow in autumn. I shall never forget them in Dublin, New Hampshire, driving through what our delightful Yankee charioteer and guide called "only a cat-road."

This was to me a new use of the word cat as a prænomen, though I knew, as did Dr. Holmes and Hosea Biglow, and every good New Englander, that "cat-sticks" were poor spindling sticks, either growing or in a load of cut wood. I heard a country parson say as he regarded ruefully a gift of a sled load of firewood, "The deacon's load is all cat-sticks." Of course a cat-stick was also the stick used in the game of ball called tip-cat. Myself when young did much practise another loved ball game, "one old cat," a local favorite, perhaps a local name. "Cat-ice," too, is a good old New England word and thing; it is the thin layer of brittle ice formed over puddles, from under which the water has afterward receded. If there lives a New Englander too old or too hurried to rejoice in stepping upon and crackling the first "cat-ice" on a late autumn morning, then he is a man; for no New England girl, a century old, could be thus indifferent. It is akin to rustling through the deep-lying autumn leaves, which affords a pleasure so absurdly disproportioned and inexplicable that it is almost mysterious. Some of us gouty ones, alas! have had to give up the "cat-slides" which were also such a delight; the little stretches of glare ice to which we ran a few steps and slid rapidly over with the impetus. But I must not let my New England folk-words lure me away from my subject, even on a tempting "cat-slide."

Though garden flowers run everywhere that they will, they are not easily forced to become wild flowers. We hear much of the pleasure of sowing garden seeds along the roadside, and children are urged to make beautiful wild gardens to be the delight of passers-by. Alphonse Karr wrote most charmingly of such sowings, and he pictured the delight and surprise of country folk in the future when they found the choice blooms, and the confusion of learned botanists in years to come. The delight and surprise and confusion would have been if any of his seeds sprouted and lived! A few years ago a kindly member of our United States Congress sent to me from the vast seed stores of our national Agricultural Department, thousands of packages of seeds of common garden flowers to be given to the poor children in public kindergartens and primary schools in our great city. The seeds were given to hundreds of eager flower lovers, but starch boxes and old tubs and flower pots formed the limited gardens of those Irish and Italian children, and the Government had sent to me such "hats full, sacks full, bushel-bags full," that I was left with an embarrassment of riches. I sent them to Narragansett and amused myself thereafter by sowing several pecks of garden seeds along the country roadsides; never, to my knowledge, did one seed live and produce a plant. I watched eagerly for certain plantings of Poppies, Candytuft, Morning-glories, and even the indomitable Portulaca; not one appeared. I don't know why I should think I could improve on nature; for I drove through that road yesterday and it was radiant with Wild Rose bloom, white Elder, and Meadow Beauty; a combination that Thoreau thought and that I think could not be excelled in a cultivated garden. Above all, these are the right things in the right place, which my garden plants would not have been. I am sure

Overgrown Garden at Llanerck, Pennsylvania.

that if they had lived and crowded out these exquisite wild flowers I should have been sorry enough.

The hardy Colchicum or Autumnal Crocus is seldom seen in our gardens; nor do I care for its increase, even when planted in the grass. It bears to me none of the delight which accompanies the spring Crocus, but seems to be out of keeping with the

Fountain at Yaddo.

autumnal season. Rising bare of leaves, it has but a seminatural aspect, as if it had been stuck rootless in the ground like the leafless, stemless blooms of a child's posy bed. Its English name — Naked Boys — seems suited to it. The Colchicum is associated in my mind with the Indian Pipe and similar growths; it is curious, but it isn't pleasing. As the Indian Pipe could not be lured within garden walls, I will not write of it here, save to say that no one could ever see it growing in its shadowy home in the woods without yielding to its air of

Avenue of White Pines at Wellesley, Mass., the Country-seat of Hollis H. Hunnewell, Esq.

mystery. It is the weirdest flower that grows, so palpably ghastly that we feel almost a cheerful satisfaction in the perfection of its performance and our own responsive thrill, just as we do in a good ghost story.

Many wild flowers which we have transplanted to our gardens are full of magic and charm. In some, such as Thyme and Elder, these elements come from English tradition. In other flowers the quality of mystery is inherent. In childhood I absolutely abhorred Bloodroot; it seemed to me a fearsome thing when first I picked it. I remember well my dismay, it was so pure, so sleek, so innocent of face, yet bleeding at a touch, like a murdered man in the Blood Ordeal.

The Trillium, Wake-robin, is a wonderful flower. I have seen it growing in a luxuriance almost beyond belief in lonely Canadian forests on the Laurentian Mountains. At this mining settlement, so remote that it was unvisited even by the omnipresent and faithful Canadian priest, was a wealth of plant growth which seemed fairly tropical. The starry flowers of the Trillium hung on long peduncles, and the two-inch diameter of the ordinary blossom was doubled. The Painted Trillium bore rich flowers of pink and wine color, and stood four or five feet from the ground. I think no one had ever gathered their blooms, for there were no women in this mining camp save a few French-Indian servants and one Irish cook, and no educated white woman had ever been within fifty, perhaps a hundred, miles of the place. Every variety of bloom seemed of exaggerated growth, but the Trillium exceeded all. An element of mystery surrounds this plant, a quality which appertains to all "three-cornered" flowers;

perhaps there may be some significance in the three-sided form. I felt this influence in the extreme when in the presence of this Canadian Trillium, so much so that I was depressed by it when wandering alone even in the edge of the forest; and when by light o' the moon I peered in on this forest garden, it was like the vision of a troop of trembling white ghosts, stimulating to the fancy. It was but a part of the whole influence of that place, which was full of eerie mystery. For after the countless eons of time during which "the earth was without form and void, and darkness was upon the face of the earth," the waters at last were gathered together and dry land appeared. And that dry land which came up slowly out of the face of the waters was this Laurentian range. And when at God's command "on the third day" the earth brought forth grass, and herb yielded seed — lo, among the things which were good and beautiful there shone forth upon the earth the first starry flowers of the white Trillium.

CHAPTER XXII

ROSES OF YESTERDAY

> "Each morn a thousand Roses brings, you say;
> Yes, but where leaves the Rose of Yesterday?"
> — *Rubaiyat of Omar Khayyam*, translated by EDWARD FITZGERALD, 1858.

HE answer can be given the Persian poet that the Rose of Yesterday leaves again in the heart. The subtle fragrance of a Rose can readily conjure in our minds a dream of summers past, and happy summers to come. Many a flower lover since Chaucer has felt as did the poet: —

> "The savour of the Roses swote
> Me smote right to the herte rote."

The old-time Roses possess most fully this hidden power. Sweetest of all was the old Cabbage Rose — called by some the Provence Rose — for its perfume "to be chronicled and chronicled, and cut and chronicled, and all-to-be-praised." Its odor is perfection; it is the standard by which I compare all other fragrances. It is not too strong nor too cloying, as are some Rose scents; it is the idealization of that distinctive sweetness of the Rose family which other Roses have to some degree. The color of the Cabbage Rose is very warm and pleasing, a clear, happy pink, and the flower has a wholesome, open look; but it is not a beautiful Rose by florists' standards, — few of the old Roses are, — and it is rather awkward in growth. The Cabbage Rose is said to have been a favorite in ancient Rome. I wish it had a prettier name; it is certainly worthy one.

The Hundred-leaved Rose was akin to the Cabbage Rose, and shared its delicious fragrance. In its rather irregular shape it resembled the present Duke of Sussex Rose.

One of the rarest of old-time Roses in our gardens to-day is the red and white mottled York and Lancaster. It is as old as the sixteenth century. Shakespeare writes in the *Sonnets*: —

> "The Roses fearfully in thorns did stand
> One blushing shame, another white despair.
> A third, nor red, nor white, had stol'n of both."

They are what Chaucer loved, "sweitie roses red, brode, and open also," Roses of a broad, flat expanse when in full bloom; they have a cheerier, heartier, more gracious look than many of the new Roses that never open far from bud, that seem so pinched and narrow. What ineffable fragrance do they pour out from every wide-open flower, a fragrance that is the very spirit of the Oriental Attar of Roses; all the sensuous sweetness of the attar is gone, and only that which is purest and best remains. I believe, in thinking of it, that it equals the perfume of the Cabbage Rose, which, ere now, I have always placed first. This York and Lancaster Rose is the *Rosa mundi*, — the rose of the world. A fine plant is growing in Hawthorne's old home in Salem.

Opposite page 462 is an unusual depiction of the century-old York and Lancaster Rose still growing

Violets in Silver Double Coaster.

and flourishing in the old garden at Van Cortlandt Manor. It is from one of the few photographs which I have ever seen which make you forgive their lack of color. The vigor, the grace, the richness of this wonderful Rose certainly are fully shown, though but in black and white. I have called this Rose bush a century old; it is doubtless much older, but it does not seem old; it is gifted with everlasting youth. We know how the Persians gather before a single plant in flower; they spread their rugs, and pray before it; and sit and meditate before it; sip sherbet, play the lute and guitar in the moonlight; bring their friends and stand as in a vision, then talk in praises of it, and then all serenade it with an ode from Hafiz and depart. So would I gather my friends around this lovely old Rose, and share its beauty just as my friends at the manor-house share it with me; and as the Persians, we would praise it in sunlight and by moonlight, and sing its beauty in verses. This York and Lancaster Rose was known to Parkinson in his day; it is his *Rosa versicolor*. I wonder why so few modern gardens contain this treasure. I know it does not rise to all the standards of the modern Rose growers; but it possesses something better — it has a living spirit; it speaks of history, romance, sentiment; it awakens inspiration and thought, it has an ever living interest, a significance. I wonder whether a hundred years from now any one will stand before some Crimson Rambler, which will then be ancient, and feel as I do before this York and Lancaster goddess.

The fragrance of the sweetest Roses — the Damask, the Cabbage, the York and Lancaster — is beyond any other flower-scent, it is irresistible, enthralling; you cannot leave it. You can push aside a Syringa, a Honeysuckle, even a Mignonette, but there is a magic something which binds you irrevocably to the Rose. I have never doubted that the Rose has some compelling quality shared not by

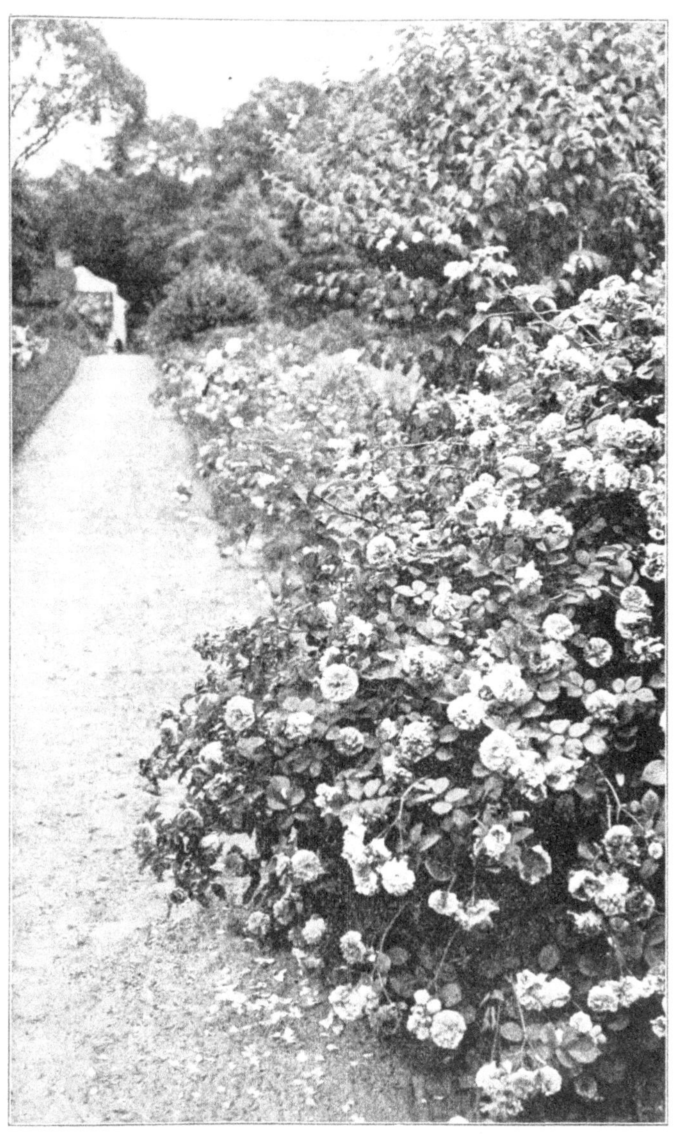

York and Lancaster Rose.

other flowers. I know not whether it comes from centuries of establishment as a race-symbol, or from some inherent witchery of the plant, but it certainly exists.

The variety of Roses known to old American gardens, as to English gardens, was few. The English Eglantine was quickly established here in gardens and spread to roadsides. The small, ragged, cheerful little Cinnamon Rose, now chiefly seen as a garden stray, is undoubtedly old. This Rose diffuses its faint "sinamon smelle" when the petals are dried. Nearly all of the Roses vaguely thought to be one or two hundred years old date only, within our ken, to the earlier years of the nineteenth century. The Seven Sisters Rose, imagined by the owner of many a Southern garden to belong to colonial days, is one of the family *Rosa multiflora*, introduced from Japan to England by Thunberg. Its catalogue name is Greville. I think the Seven Sisters dates back to 1822. The clusters of little double blooms of the Seven Sisters are not among our beautiful Roses, but are planted by the house mistress of every Southern home from power of association, because they were loved by her grandmothers, if not by more distant forbears. The crimson Boursaults are no older. They came from the Swiss Alps and therefore are hardy, but they are fussy things, needing much pruning and pulling out. I recall that they had much longer prickles than the other roses in our garden. The beloved little Banksia Rose came from China in 1807. The Madame Plantier is a hybrid China Rose of much popularity. We have had it about seventy or eighty years. In the lovely garden of Mrs. Mabel Osgood Wright, author of *Flowers and Trees in their Haunts*, I saw, this spring, a giant Madame Plantier which had over five thousand buds, and which could scarcely be equalled in beauty by any modern Roses. Its photograph gives scant idea of its size.

What gratitude we have in spring to the Sweetbrier! How early in the year, from sprouting branch and curling leaf, it begins to give forth its pure odor! Gracious and lavish plant, beloved in scent by every one, you have no rival in the spring garden with its pale perfumes. The Sweetbrier and Shakespeare's Musk Rose (*Rosa moschata*) are said to be the only Roses that at evening pour forth their perfume; the others are what Bacon called "fast of their odor."

The June Rose, called by many the Hedgehog Rose, was, I think, the first Rose of summer. A sturdy plant, about three feet in height; set thick with briers, it well deserved its folk name. The flowers opened into a saucer of richest carmine, as fragrant as an American Beauty, and the little circles of crimson resembling the *Rosa rugosa* were seen in every front dooryard.

In the Walpole garden from whence came to us our beloved Ambrosia, was an ample Box-edged flower bed which my mother and the great-aunt called The Rosery. One cousin, now living, recalls with distinctness its charms in 1830; for it was beautiful, though the vast riches of the Rose-world of China and Japan had not reached it. There grew in it, he remembers, Yellow Scotch Roses, Sweetbrier (or Eglantine), Cinnamon Roses, White Scotch Roses, Damask Roses, Blush Roses, Dog Roses (the Canker-bloom of Shakespeare), Black Roses, Burgundy Roses, and Moss Roses. The last-named sensitive creatures, so difficult to rear with satisfaction in such a climate, found in this Rosery by the

Cinnamon Roses.

river-side some exact fitness of soil or surroundings, or perhaps of fostering care, which in spite of the dampness and the constant tendency of all Moss Roses to mildew, made them blossom in unrivalled perfection. I remember their successors, deplored as much inferior to the Roses of 1830, and they were the finest Moss Roses I ever saw blooming in a garden. An amusing saying of some of the village passers-by (with smaller gardens and education) showed the universal acknowledgment of the perfection of these Roses. These people thought the name was Morse Roses and always thus termed them, fancying they were named for the family for whom the flowers bloomed in such beauty and number.

Among the other Roses named by my cousin I recall the White Scotch Rose, sometimes called also the Burnet-leaved Rose. It was very fragrant, and was often chosen for a Sunday posy. There were both single and double varieties.

The Blush Rose (*Rosa alba*), known also as Maiden's blush, was much esteemed for its exquisite color; it could be distinguished readily by the glaucous hue of the foliage, which always looked like the leaves of artificial roses. It was easily blighted; and indeed we must acknowledge that few of the old Roses were as certain as their sturdy descendants.

The Damask Rose was the only one ever used in careful families and by careful housekeepers for making rose-water. There was a Velvet Rose, darker than the Damask and low-growing, evidently the same Rose. Both showed plentiful yellow stamens in the centres, and had exquisite rich dark leaves.

The old Black Rose of The Rosery was so suffused with color-principle, so "color-flushing," that even the wood had black and dark red streaks. Its petals were purple-black.

The Burgundy Rose was of the Cabbage Rose family; its flowers were very small, scarce an inch in diameter. There were two varieties: the one my cousin called Little Burgundy had clear dark red blossoms; the other, white with pink centres. Both were low-growing, small bushes with small leaves. They are practically vanished Roses—wholly out of cultivation.

We had other tiny roses; one was a lovely little Rose creature called a Fairy Rose. I haven't seen one for years. As I recall them, the Rose plants were never a foot in height, and had dainty little flower rosettes from a quarter to half an inch in diameter set in thick clusters. But the recalled dimensions of youth vary so when seen actually in the cold light of to-day that perhaps I am wrong in my description. This was also called a Pony Rose. This Fairy Rose was not the Polyantha which also has forty or fifty little roses in a cluster. The single Polyantha Rose looks much like its cousin, the Blackberry blossom.

Another small Rose was the Garland Rose. This was deemed extremely elegant, and rightfully so. It has great corymbs of tiny white blossoms with tight little buff buds squeezing out among the open Roses.

Another old favorite was the Rose of Four Seasons—known also by its French name, *Rose de Quartre Saisons*—which had occasional blooms throughout the summer. It may have been the foundation of our Hybrid Perpetual Roses. The Bourbon Roses were vastly modish; their round smooth petals and oval leaves easily distinguish them from other varieties.

Among the several hundred things I have fully planned out to do, to solace my old age after I have become a "centurion," is a series of water-color drawings of all these old-time Roses, for so many of them are already scarce.

The Michigan Rose which covered the arches in Mr. Seward's garden, has clusters of deep pink, single, odorless flowers, that fade out nearly white after they open. It is our only native Rose that has passed into cultivation. From it come many fine double-flowered Roses, among them the beautiful Baltimore Belle and Queen of the Prairies, which were named about 1836 by a Baltimore florist called Feast. All its vigorous and hardy descendants are scentless save the Gem of the Prairies. It is one of the ironies of plant-nomenclature when we have so few plant names saved to us from the picturesque and often musical speech of the American Indians, that the lovely Cherokee Rose, Indian of name, is a Chinese Rose. It ought to be a native, for everywhere throughout our Southern states its pure white flowers and glossy evergreen leaves love to grow till they form dense thickets.

People who own fine gardens are nowadays unwilling to plant the old "Summer Roses" which bloom cheerfully in their own Rose-month and then have no more blossoming till the next year; they want a Remontant Rose, which will bloom a second time in the autumn, or a Perpetual Rose, which will give flowers from June till cut off by the frost. But these latter-named Roses are not only of fine gardens but of fine gardeners; and folk who wish the old

Cottage Garden with Roses.

simple flower garden which needs no highly-skilled care, still are happy in the old Summer Roses I have named.

A Rose hedge is the most beautiful of all garden walls and the most ancient. Professor Koch says that long before men customarily surrounded their gardens with walls, that they had Rose hedges. He tells us that each of the four great peoples of Asia owned its own beloved Rose, carried in all wanderings, until at last the four became common to all races of men. Indo-Germanic stock chose the hundred-leaved red Rose, *Rosa gallica* (the best Rose for conserves). *Rosa damascena*, which blooms twice a year, and the Musk Rose were cherished by the Semitic people; these were preferred for attar of Roses and Rose water. The yellow Rose, *Rosa lutea*, or Persian Rose, was the flower of the Turkish Mongolian people. Eastern Asia is the fatherland of the Indian and Tea Roses. The Rose has now become as universal as sunlight. Even in Iceland and Lapland grows the lovely *Rosa nitida*.

We say these Roses are common to all peoples, but we have never in America been able to grow yellow Roses in ample bloom in our gardens. Many that thrive in English gardens are unknown here. The only yellow garden Rose common in old gardens was known simply as the "old yellow Rose," or Scotch Rose, but it came from the far East. In a few localities the yellow Eglantine was seen.

The picturesque old custom of paying a Rose for

rent was known here. In Manheim, Pennsylvania, stands the Zion Lutheran Church, which was gathered together by Baron William Stiegel, who was the first glass and iron manufacturer of note in this country. He came to America in 1750, with a fortune which would be equal to-day to a million dollars, and founded and built and named Manheim. He was a man of deep spiritual and religious belief, and of profound sentiment, and when in 1771 he gave the land to the church, this clause was in the indenture: —

> "Yielding and paying therefor unto the said Henry William Stiegel, his heirs or assigns, at the said town of Manheim, in the Month of June Yearly, forever hereafter, the rent of *One Red Rose*, if the same shall be lawfully demanded."

Nothing more touching can be imagined than the fulfilment each year of this beautiful and symbolic ceremony of payment. The little town is rich in Roses, and these are gathered freely for the church service, when One Red Rose is still paid to the heirs of the sainted old baron, who died in 1778, broken in health and fortunes, even having languished in jail some time for debt. A new church was erected on the site of the old one in 1892, and in a beautiful memorial window the decoration of the Red Rose commemorates the sentiment of its benefactor.

The Rose Tavern, in the neighboring town of Bethlehem, stands on land granted for the site of a tavern by William Penn, for the yearly rental of One Red Rose.

In England the payment of a Rose as rent was often known. The Bishop of Ely leased Ely house in 1576 to Sir Christopher Hatton, Queen Elizabeth's handsome Lord Chancellor, for a Red Rose to be paid on Midsummer Day, ten loads of hay and ten pounds per annum, and he and his Episcopal successors reserved the right of walking in the gardens and gathering twenty bushels of Roses yearly. In France there was a feudal right to demand a payment of Roses for the making of Rose water.

Two of our great historians, George Bancroft and Francis Parkman, were great rose-growers and rose-lovers. I never saw Mr. Parkman's Rose Garden, but I remember Mr. Bancroft's well; the Tea Roses were especially beautiful. Mr. Bancroft's Rose Garden in its earliest days had no rivals in America.

The making of potpourri was common in my childhood. While the petals of the Cabbage Rose were preferred, all were used. Recipes for making potpourri exist in great number; I have seen several in manuscript in old recipe books, one dated 1690. The old ones are much simpler than the modern ones, and have no strong spices such as cinnamon and clove, and no bergamot or mints or strongly scented essences or leaves. The best rules gave ambergris as one of the ingredients; this is not really a perfume, but gives the potpourri its staying power. There is something very pleasant in opening an old China jar to find it filled with potpourri, even if the scent has wholly faded. It tells a story of a day when people had time for such things. I read in a letter a century and a half old of a happy group of people riding out to the house of the provincial governor of New York; all gathered Rose leaves in the governor's garden, and the governor's wife started the distilling of these Rose leaves, in her new still, into Rose water, while all drank syllabubs and junkets — a pretty Watteau-ish scene.

The hips of wild Roses are a harvest — one unused in America in modern days, but in olden times they were stewed with sugar and spices, as were other fruits. Sauce Saracen, or Sarzyn, was made of Rose hips and Almonds pounded together, cooked in wine and sweetened. I believe they are still cooked by some folks in England, but I never heard of their use in America save by one person, an elderly Irish woman on a farm in Narragansett. Plentiful are the references and rules in old cookbooks for cooking Rose hips. Parkinson says: "Hippes are made into a conserve, also a paste like licoris. Cooks and their Mistresses know how to prepare from them many fine dishes for the Table." Gerarde writes characteristically of the Sweetbrier, "The fruit when it is ripe maketh most pleasant meats and banqueting dishes, as tarts and such-like; the making whereof I commit to the cunning cooke, and teeth to eat them in the rich man's mouth."

Children have ever nibbled Rose hips: —

"I fed on scarlet hips and stony haws —
Hard fare, but such as boyish appetite
Disdains not."

The Rose bush furnished another comestible for the children's larder, the red succulent shoots of common garden and wild Roses. These were known by the dainty name of "brier candy," a name appropriate and characteristic, as the folk-names devised by children frequently are.

On the post-road in southern New Hampshire stands an old house, which according to its license was once "improved" as a tavern, and was famous for its ghost and its Roses. The tavern was owned by a family of two brothers and two sisters, all unmarried, as was rather a habit in the Mason family; though when any of the tribe did marry, a vast throng of children quickly sprung up to propagate the name and sturdy qualities of the race. The men were giants, and both men and women were hard-working folk of vast endurance and great thrift, and, like all of that ilk in New England, they prospered and grew well-to-do; great barns and outbuildings, all well filled, stretched down along the roadside below the house. Joseph Mason could lay more feet of stone wall in a day, could plough more land, chop down more trees, pull more stumps, than any other man in New Hampshire. His sisters could bake and brew, make soap, weed the garden, spin and weave, unceasingly and untiringly. Their garden was a source of purest pleasure to them, as well as of hard work; its borders were so stocked with medicinal herbs that it could supply a township; and its old-time flowers furnished seeds and slips and bulbs to every other garden within a day's driving distance; but its glory was a garden side to gladden the heart of Omar Khayyam, where two or three acres of ground were grown over heavily with old-fashioned Roses. These were only the common Cinnamon Rose, the beloved Cabbage Rose, and a pale pink, spicily scented, large-petalled, scarcely double Rose, known to them as the Apothecaries'

Madame Plantier Rose.

Rose. Farmer-neighbors wondered at this waste of the Masons' good land in this unprofitable Rose crop, but it had a certain use. There came every June to this Rose garden all the children of the vicinity, bearing milk-pails, homespun bags, birch baskets, to gather Rose petals. They nearly all had Roses at their homes, but not the Mason

Roses. These Rose leaves were carried carefully to each home, and were packed in stone jars with alternate layers of brown or scant maple sugar. Soon all conglomerated into a gummy, brown, close-grained, not over alluring substance to the vision, which was known among the children by the unromantic name of "Rose tobacco." This cloying confection was in high repute. It was chipped off and eaten in tiny bits, and much treasured — as a love token, or reward of good behavior.

The Mason house was a tavern. It was not one of the regular stopping-places on the turnpike road, being rather too near the town to gather any travel of teamsters or coaches; but passers-by who knew the house and the Masons loved to stop there. Everything in the well-kept, well-filled house and barns contributed to the comfort of guests, and it was known that the Masons cared more for the company of the traveller than for his pay.

There was a shadow on this house. The youngest of the family, Hannah, had been jilted in her youth, "shabbed" as said the country folks. After several years of "constant company-keeping" with the son of a neighbor, during which time many a linen sheet and tablecloth, many a fine blanket, had been spun and woven, and laid aside with the tacit understanding that it was part of her wedding outfit, the man had fallen suddenly and violently in love with a girl who came from a neighboring town to sing a single Sunday in the church choir. He had driven to her home the following week, carried her off to a parson in a third town, married her, and brought her to his home in a triumph of enthusiasm and romance, which quickly fled before the open dislike and reprehension of his upright neighbors, who abhorred his fickleness, and before the years of ill health and ill temper of the hard-worked, faded wife. Many children were born to them; two lived, sickly little souls, who, unconscious of the blemish on their parents' past, came with the other children every June, and gathered Rose leaves under Hannah Mason's window.

Hannah Mason was called crazy. After her desertion she never entered any door save that of her own home, never went to a neighbor's house either in time of joy or sorrow; queerer still, never went to church. All her life, her thoughts, her vast strength, went into hard work. No labor was too heavy or too formidable for her. She would hetchel flax for weeks, spin unceasingly, and weave on a hand loom, most wearing of women's work, without thought of rest. No single household could supply work for such an untiring machine, especially when all labored industriously — so work was brought to her from the neighbors. Not a wedding outfit for miles around was complete without one of Hannah Mason's fine tablecloths. Every corpse was buried in one of her linen shrouds. Sailmakers and boat-owners in Portsmouth sent up to her for strong duck for their sails. Lads went up to Dartmouth College in suits of her homespun. Many a teamster on the road slept under Hannah Mason's heavy gray woollen blankets, and his wagon tilts were covered with her canvas. Her bank account grew rapidly

Sun-dial and Roses at Van Cortlandt Manor.

— she became rich as fast as her old lover became poor. But all this cast a shadow on the house. Sojourners would waken and hear throughout the night some steady sound, a scratching of the cards, a whirring of the spinning-wheel, the thump-thump of the loom. Some said she never slept, and could well grow rich when she worked all night.

At last the woman who had stolen her lover — the poor, sickly wife — died. The widower, burdened hopelessly with debts, of course put up in her memory a fine headstone extolling her virtues. One wakeful night, with a sentiment often found in such natures, he went to the graveyard to view his proud but unpaid-for possession. The grass deadened his footsteps, and not till he reached the grave did there rise up from the ground a tall, ghostly figure dressed all in undyed gray wool of her own weaving. It was Hannah Mason. "Hannah," whimpered the widower, trying to take her hand,—with equal thought of her long bank account and his unpaid-for headstone, — "I never really loved any one but you." She broke away from him with an indescribable gesture of contempt and dignity, and went home. She died suddenly four days later of pneumonia, either from the shock or the damp midnight chill of the graveyard.

As months passed on travellers still came to the tavern, and the story began to be whispered from one to another that the house was haunted by the ghost of Hannah Mason. Strange sounds were heard at night from the garret where she had always worked; most plainly of all could be heard the whirring of her great wool wheel. When this rumor reached the brothers' ears, they determined to investigate the story and end it forever. That night their vigil began, and soon the sound of the wheel was heard. They entered the garret, and to their surprise found the wheel spinning round. Then Joseph Mason went to the garret and seated himself for closer and more determined watch. He sat in the dark till the wheel began to revolve, then struck a sudden light and found the ghost. A great rat had run out on the spoke of the wheel and when he reached the broad rim had started a treadmill of his own — which made the ghostly sound as it whirred around. Soon this rat grew so tame that he would come out on the spinning-wheel in the daytime, and several others were seen to run around in the wheel as if it were a pleasant recreation.

The old brick house still stands with its great grove of Sugar Maples, but it is silent, for the Masons all sleep in the graveyard behind the church high up on the hillside; no travellers stop within the doors, the ghost rats are dead, the spinning-wheel is gone, but the garden still blossoms with eternal youth. Though children no longer gather rose leaves for Rose tobacco, the "Roses of Yesterday" bloom every year; and each June morn, "a thousand blossoms with the day awake," and fling their spicy fragrance on the air.

Index

Printed in the United States
By Bookmasters